Science of Electricity

Volume 10

Utility Operations and Grid Systems
Explained Simply

by Mark Fennell
© 2012

This book is part of the
Energy Technologies Explained Simply™ Series

Other Books in the Energy Technology Series

About the Book

<u>Overview</u>

After reading this book you will understand everything you need to know about utility and grid operations. Utilities and power grids may seem complex, yet this book will guide you easily through the mysteries.

In this book you will learn the basic operations of utilities and the basic operations of grids. You will learn how power is traded in the power markets. You will learn how utilities and grids maintain quality control, including how they monitor the flow of power and how they make adjustments as needed. You will learn how power failures occur and how we can minimize power failures in the future.

You will also learn a great deal about the future of power distribution, particularly in the areas of distributed generation and smart grids.

No technical background is required; this book will explain everything you need to know about the operation of utilities and power grids in a way that any reader can understand.

This book is designed for policy-makers, community activists, and curious citizens. This book is also designed as an essential reference for energy technology students and for anyone working in the electrical power industry.

10.1 <u>Utility Company Operation Basics</u>

The first chapter provides an overview of utility operation, focusing on quality control for utilities. This chapter discusses in detail how to ensure that there is enough power to meet demand at any given time.

10.2 <u>Monitoring and Communications Systems</u>

The second chapter discusses monitoring and communication systems. This chapter explains the SCADA system, including the components and operation. This chapter also explains the types of translation devices and the options for communication methods.

10.3 Quality Control for Utilities

Chapter three discusses the specific factors which affect the quality of power (including voltage, frequency, and temperature). The majority of this chapter describes the causes, effects, and protection against significant variations for each factor.

10.4 Basic Concepts of Grids

Chapter four provides a broad perspective of grids, with further clarification by comparing grids to the highway system. You will learn the possible players in a grid and their roles. You will learn the jobs of the grid manager. A major section of this chapter is where you will learn the advantages and disadvantages of grids, followed by an overview of an ideal grid system.

10.5 Grid Operations

In chapter five we go step by step through the sequence of grid operations. By the end of this chapter you will understand the details of grid operations, including several common variations.

10.6 Quality Control for Grids

Chapter six discusses grid failures and quality control for grids. In this chapter you will learn how blackouts occur and how we can prevent them. The majority of chapter six provides detailed explanations of how to maintain quality control in any grid system, and thereby prevent power outages.

You will learn about reliability oversight organizations such as NERC, FERC, and ERO. You will also learn the major sections of the Energy Policy Act of 2005 (most of which relate to reliability).

10.7 Smart Grids

Chapter seven discusses "Smart Grids". There are numerous technologies associated with the term "smart grid", therefore chapter seven begins with an overview of smart grids, followed by an overview of smart grid technologies. Also, because smart grids are rapidly implemented yet often used in unethical ways, much of this chapter is devoted to the proper and improper uses of smart grid technologies.

10.8 The Future of Electrical Distribution

The final chapter discusses the future of electrical distribution. This is a very important chapter because the paths we take when developing power distribution systems will affect our safety, reliability, and economic security for many years.

After years of research I have learned that the existing trend toward a national grid is a direct path to large scale disasters. I have also learned that an alternate path, one of locally produced power and smaller power plants, will provide reliable power and a stable economy for decades.

Therefore, this chapter explains the vision of this second path for the future of electrical distribution. The description of this vision is then supplemented with benefits of the second path. Each benefit of the second path is discussed in great detail. This chapter also discusses some additional features which are valuable for the ideal future of electrical distribution.

About the Energy Technology Series

Purpose of this series

The books in the *Energy Technologies* series are designed to educate citizens, students, and legislators on all aspects of energy technologies. The first books in the series focus on electrical power.

The books discuss many energy technologies, including: generators, turbines, power plants, power lines, and grids. The technologies for each type of power source are discussed in detail. The books also discuss efficiency, reliability, and health concerns for each energy technology.

The ultimate goal of the series is to enable the people to make informed decisions on practical energy questions. The secondary goal is to serve as introductory guides for students embarking on careers with energy technologies.

Taken altogether the books in the series answer any question you are likely to have, such as:

- How can we increase the efficiency of solar cells?
- What do I need to know when installing a wind turbine?
- How effective are the clean coal technologies?
- How can we prevent grid failures?
- Do power lines cause cancer?
- and many other energy technology questions...

Science of Electricity in Perspective

The subject of electrical power is of great importance to our communities, but is rarely taught. Public debate is frequent and passionate, but with too little understanding of the actual science. At best, an informed citizen knows only a few pieces. At worst, as it is for a great number of citizens, electricity is magic and myths are believed as scientific truth. It does not have to be that way. Any citizen, regardless of background, can know the technologies behind all aspects of electricity.

The books in this series solve that problem. These books educate the general public in all aspects of electrical power. Any person, regardless of background, can easily find the answer to his energy question in one of these books.

Specific Goals

There are numerous technologies described in these books. Yet for each technology I sought out the answers to the following questions:

1. How does the technology work?
2. What are the advantages and disadvantages?
3. What is the efficiency? How can the efficiency be improved?
4. What is the environmental impact? How can it be improved?
5. What are the safety hazards, and how can they be reduced?
6. What are the most important practical tips?
7. What facts comprise the most important data?

Technical Discussions Explained Simply

The books in the series must necessarily be technical to some degree. Electricity is a practical technology, and therefore we must understand the technical aspects if we want to make wise decisions. Yet the discussions in this book are always aimed at the citizen or policy maker.

The books in this series explain the principles of electricity as simply as possible, using ordinary English (no engineering jargon), and highlighting the most important points of each technology. Main concepts and facts are emphasized with the use of lists, tables, diagrams, and summaries.

I do not expect any reader to have a background in science, yet I offer enough facts and details so that the reader can have an accurate understanding of all related technologies. I provide enough technical details and enough data for the reader to make informed decisions.

Accuracy of Data and Summaries

I never relied solely on the conclusions of other researchers. Instead, I checked the accuracy of the conclusions written by other researchers, most commonly by finding at least three distinct sources for each fact. I have also performed my own research and talked with experts in various subjects. It is only after such rigorous investigations that I created data tables and wrote summaries for these books.

<u>Objectivity</u>

Throughout the series of books, I have tried my best to be as objective as possible. Whereas many other authors of energy books have an agenda, I have no desire to promote one industry over another. I have no desire to promote one technical solution over another. In this endeavor, I have tried to be an objective scientist.

M.F.

Passion Behind The Future of Grids

Note that this book differs from all other books in the series because I will be openly passionate about a particular topic. In this case my passion is related to the future of grids.

Grids affect all of us, and in a very large way. Unlike a single power plant or a local utility where the choice affects only a few people, what happens on a large grid will affect all of us. Therefore the design of a grid matters greatly.

After years of research, travels, and discussions, I have arrived at some definite conclusions regarding grids. There are two clear paths. The first path is the one we are on. It leads to large scale power outages, unreliable power, economic uncertainty, and chaos. The other path will provide reliability, safety, and economic security...and this second path will provide those benefits for decades. This second choice is clearly the better path.

And yet the few experts who talk about this better path are not being heard. Therefore this is an issue on which I cannot be complacent. We need to be on this better path, moving toward this better future of grid systems. I encourage policy-makers and citizens to move toward this second path. On this issue, I cannot remain unbiased.

M.F.

Table of Contents

Table of Contents: Detailed

10.1
Utility Company Operation Basics

Introduction

Effective operation of an electrical company means reliability. In practical terms, reliability means that customers can count on their service without major interruptions. If a utility company wants to provide reliable electricity, then the utility company must be able to do several specific tasks. These tasks are:

- Match power generation with load
- Create enough reserve
- Build quality electrical lines and distribution systems
- Know the limitations of lines and equipment
- Have an advanced monitoring and communications system
- Switch transmission paths quickly when any piece fails
- Disconnect and repair equipment quickly
- Restart a power plant without difficulty

List of topics for this chapter

1. Overview of Matching Power Generation with Load
2. Problems if Generation does Not Meet Load
3. Methods for Matching Generation with Load
4. Create Enough Reserve
5. Build Quality Electrical Lines and Distribution Systems
6. Monitor Key Factors and Respond to Changes in the System
7. Have an Effective Monitoring and Communications System
8. Restarting Power Plants Without Difficulty
9. Interconnect with Other Power Systems Effectively

Match Generation with Load: Overview

Introduction

When utility companies talk of matching generation with load, they are simply referring to matching supply with demand. The demand for electricity is the amount of power that the people, collectively, draw from the system at any given time. This changes moment by moment. The utility company must then figure a way to meet that demand. If the utility company can't do this, then someone will not have power.

Demand vs. Load

The terms "demand" and "load" are similar, but there is a subtle distinction. This distinction is important whenever we discuss subjects such as utility companies or grids. "Demand" is the amount of power which the people use at any given time. "Load" is the amount of power transmitted over the wire. Ideally, the demand and the load should be the same. However, there will always be some power loss along the way. Therefore, the "load" must always be greater than the "demand." Stated another way: Load = Demand + Expected Power Loss.

Problems if Generation Does Not meet Load

Introduction

If generation does not meet demand, then one or more of the following might happen:

1. Frequency variation (which could damage power plants)
2. Damage to transmission equipment
3. Power lost to local consumers
4. Power supply variations to other connected power systems

1. Frequency Variation (which could damage power plants)

When there is not enough supply to meet demand the frequency of the electricity may decrease. A decline in frequency can be very bad. If the frequency drops by just a small amount, such as a 5 Hertz difference, then a power plant may have a complete loss of power. This is particularly true for steam operated plants such as nuclear power and coal power plants.

2. Damage to Transmission Equipment

Drops in frequency or voltage can seriously damage some types of transmission equipment. If these pieces of equipment become damaged, then many people will lose their power. Furthermore, these pieces of equipment often require a long time to repair or restart.

3. Power Lost to Local Consumers

This is the most direct consequence. If supply does not meet demand, then some customers will lose their power.

4. Power Supply to Other Connected Power Systems Altered

We will discuss this issue more when we discuss interconnected power systems. Briefly, if one company promised to give electricity to another utility company, but cannot deliver at the moment because of its own temporary inability to generate enough power, then both power companies are affected. People in both communities may have durations of low power, or even complete loss of power.

Methods for Matching Generation with Load

Introduction

The most common methods for meeting the load are:

1. Anticipate demand
2. Get power from other power plants
3. Have enough reserve power
4. Load shedding (dropping service to customers one by one)

Anticipate Demand

Utilities constantly try to anticipate the demand. However, this is not an easy task. The demand for power fluctuates, depending on the time of day, the season, and various activities in the area. Furthermore, in order to generate electricity for a particular time, a generating facility must start producing hours or days earlier.

Like other industries, utility companies watch the trends and plan accordingly. They create power or buy power ahead of the time needed, and they schedule delivery of that power for specific hours. Yet with all this planning, the actual demand usually varies from the anticipated amount.

Getting Power from Other Power Plants

It is very common today for utility companies to get power from other power plants. There are generally two methods. Some energy companies have their own additional power plants which remain idle until needed. These power plants are used only during peak electricity use.

In addition, many utilities buy their additional power from other energy companies. This power is bought and delivered through the regional grid system. Grids will be discussed in detail in later chapters.

Reserve Power

Reserve power is a secondary source of power which can be called upon quickly when actual demand exceeds the anticipated value. Common power reserves include natural gas and spinning reserves. Reserve power will be discussed in detail below.

Load Shedding

Load shedding is simple enough. Power to customers is dropped one customer at a time. With load shedding, only a few customers will be without power rather than many people. Power will be restored to customers as supply becomes available. This needs to happen quickly. Therefore, load shedding must be planned long before it needs to be implemented.

Some utilities have asked large consumers to sign up for voluntary load shedding. When the time comes, utilities know they can start dropping power of those customers first. In return, those businesses get discounts in their monthly bill. Many utilities have computer programs which estimate where load shedding should be done. The criteria used by the program must be determined by the specific utility company. Note that in some regions, load shedding is also known as "rolling blackouts."

Create Enough Reserve

Introduction

One of the methods for matching generation with load, as listed above, is to have enough reserve. However, creating and using reserves can be tricky because electricity is difficult to store. The most common types of reserves are:

1. Batteries
2. Natural Gas and Diesel
3. Water Reservoirs (for hydropower)
4. Spinning Reserves

Batteries

Batteries are the simplest form of electrical reserves. However, they are the least practical for utility companies. If an individual produces his own power, such as with solar or wind, then batteries can be effective as electrical reserves. Similarly, if a business produces its own power to meet its electrical needs then batteries can be effective as electrical reserves. However, for a utility company which generates power for millions of customers it would not be practical to store electricity for all those customers in a system of batteries.

Natural Gas and Diesel

Natural gas and diesel are both effective as electrical reserves. These fuels store well for long periods of time, and yet these fuels can reach full power relatively quickly.

Utility companies often have a secondary plant, used only during peak use times. These secondary plants are usually natural gas. Fuel is stored on-site, and when more electricity is required the fuel is burned in a combustion turbine.

Diesel gasoline generators can also be used to provide supplemental electricity. However, diesel generators are more effectively used as back-up generators when the original power source fails, rather than being used as peak-use generators.

Hydropower Systems

Hydropower plants are effective for providing additional power when needed. There are several ways to use hydropower for this purpose.

Many large utility companies use hydropower during peak use times by opening up additional sluice gates, which allows additional power to be generated. In addition, some utilities have smaller hydro plants used only during peak use times.

Hydropower plants can be built in almost any size. Therefore small hydropower plants can be built at many locations. These smaller hydro systems often sit behind the primary power plant or on the land of a large business. The water for these hydropower plants may come from rainfall as well as from local rivers. When supplemental power is needed, the facility can access this small hydro system to provide the power.

Another method is to have a contained system of water, where the water is pumped up to the reservoir for storage. For example, at a particular location wind turbines provide all regional power. Most of the power generated is sent directly to the customers, yet some of that power is used to pump water up to the reservoir. Later, when the wind is not blowing the facility managers will start the hydro system. This hydro system has a full reservoir and will now provide the power the community. When the wind turbines are operating again, some of this power will again be used to pump the water back to the reservoir.

Spinning Reserve

One of the most common types of electrical reserves used by utility companies is the spinning reserve. It is the spinning reserve which we will discuss in greatest detail.

The spinning reserve is a live reserve that the power plants try to keep on hand, in case the people pull more power than usual. The spinning reserve is extra power created by the generators at the power plant. This extra power is created at the same time as the scheduled power.

Stated another way, the total power generated by the power plant = scheduled power + spinning reserve.

In contrast to gas and hydropower which take time to start generating power, and in contrast to coal and nuclear which require time to increase production (even when fully operational), the spinning reserve is available immediately.

However, we must remember that the spinning reserve is a "live" reserve, and is not stored energy like a reservoir or a battery. The spinning reserve must constantly be "filled".

Note that a balance must be created. It is important that we have enough of a spinning reserve at all times in case the demand temporarily exceeds expectations. However, any unused spinning reserve is a wasted use of energy and adds to the overall cost.

Build Quality Electrical Equipment

In order to distribute the electricity to many people in an area, we must have numerous power lines, substations, and transformers. Each of these components is part of the distribution system. It is also important that we build this distribution system in a durable way. We have discussed these points previously in the chapters on transmission. (For more detail read the book "Transmission of Electrical Power").

Monitor Key Factors and Respond to Changes

Introduction

Operators must monitor and respond to changes along the power transmission system. The three key elements are:

1. Know the limits of each piece of equipment and never exceed those limits
2. Switch transmission paths when any portion of the path fails
3. Disconnect/repair damaged equipment quickly

Know the Limits of Each Piece of Equipment

The operators must know upper and lower limits for every piece of equipment, and never exceed those limits. In brief, operators must monitor voltage, frequency, and temperature. Operators must also monitor the flow of power across the grid. If any piece of the system begins to reach its limit, then the operator can take appropriate action. We will look at these concepts in greater detail in the chapters on quality control.

Switch Transmission Paths When Any Piece Fails

When one electrical line fails, then we can switch the current to go through an alternate line. When a transformer fails, we can switch the current to go through an alternate transformer. The switching is done through relays. Most of these relays are controlled remotely, and many are automatic. However, no one can predict everything, therefore human oversight is necessary during relay switching operations.

Disconnect and Repair Equipment Quickly

A good utility company will know of a problem immediately after it has occurred, and be able to fix the problem quickly after it is known. The usual process is using circuit breakers and relays which will switch automatically. The utility company then investigates.

Most problems are a very short duration: problems are simple enough to be fixed remotely. The issue is resolved within a few seconds, or at most a few minutes.

However, after a quick investigation the utility company will find some equipment that needs manual repair. This equipment is taken off-line (if not off-line already), and a utility worker is dispatched to repair the device.

Effective Monitoring & Communications Systems

It is important to monitor all aspects of electrical generation, load, and distribution. In brief, an effective monitoring and communications system must have the following:

1. Devices which turn phenomena into a signal.
2. Wires which carry the signal from remote locations to the control room.
3. Devices and wires going the other way, signaling back to the devices telling the devices what to do.
4. Phone communication with all utility workers in the field.
5. Phone communication with other power plants.
6. Backup communication, for both machines and workers.
7. Backup power supplies to power the communications.

We will discuss the details in the chapter on Monitoring and Communication Systems.

Restarting Power Plants and Grids

Introduction

It is not uncommon for power plants to shut down temporarily. Remember that circuit breakers are installed in a home in order to turn power off during a power surge or during extreme temperatures rather than damage the equipment and cause a fire. So it is with a power plant. Major equipment in a power plant will shut down automatically rather than being totally destroyed. The faults are repaired, and then the power plant is started up again.

Of course, we would rather not reach that point, and operators do their best to prevent plants from shutting down. However, it does happen. Therefore the operators need to be able to restart a power plant without too much difficulty.

We must note that restarting a power plant takes time. It is not as simple as turning on a switch, yet it usually not a serious problem. There is a set of procedures to follow for each plant, and doing those procedures will usually restart the plant without trouble. It is orderly, but it does take time. The problem is even more complex and we must be more careful when the power plant is linked to a grid. As most plants today are part of a grid, restarting a power plant can be a complex process.

Note that we must have an external source of power to restart the power plant. In other words, although the power plant exists to create power, we need a different source of power in order to start the plant up. There are several ways we can get this power:

1. Diesel gas generators
2. Batteries or pneumatic devices
3. "House" units
4. Getting power from another plant on the grid

Diesel Gas Generators

Diesel generators are the most common type of emergency generator. These come in all sizes, and are used almost everywhere. Diesel can be stored virtually forever, it provides significant amounts of energy per volume, and it converts to electrical power very quickly.

Battery or Pneumatic Devices

While the power plant is functioning, some of the power is sent to batteries to be stored. These batteries can then be used to start up the plant when needed. Pneumatic devices are essentially air pumps. Pneumatic devices can be used to start up many of the smaller pieces of equipment in the plant. However, pneumatics are not suitable for starting the main power plant itself.

House Units

House units are sources of power which automatically power certain pieces of equipment when there is trouble with the main plant. House units are generators or stored power devices which disconnect from the power plant when voltage or frequency drops. These units continue to power specific equipment of the plant for a designed time. If these house units are in place, then the power plant is easier to start up.

Getting Power from Another Plant on the Grid

This idea seems reasonable enough: get some electricity from a working plant on the grid to help start up the plant that was shut down. However, we must be careful when using the power from one plant to restart another plant. If the power is down in one plant, and is linked to another, there will be quite a drain. Therefore, although we can get power from other plants on the grid, we must do this slowly, and gradually. If we try to get power from another plant too quickly, then we will have blackouts in other areas. Note that some utility experts think this method should be done only if there are no other options.

Interconnect with Other Power Systems Effectively

The primary reason for power systems to be interconnected is to share power. If one utility needs more power, and a second utility can offer more power than it needs, then more people will have electricity at any given time. Interconnected power systems (grids) are a complex subject, and therefore will be discussed in separate chapters of this book.

Chapter Summary

1. If a utility company wants to provide reliable electricity, then the utility company must be able to do the following:
    a. Match generation with load
    b. Create enough reserve
    c. Build quality electrical lines and distribution systems
    d. Know the limitations of lines and equipment
    e. Have an advanced monitoring and communications system
    f. Switch transmission paths quickly when any piece fails
    g. Disconnect and repair equipment quickly
    h. Restart a power plant without difficulty
    i. Interconnect with other power systems effectively

2. "Demand" is the amount of power which the people use at any given time. "Load" is the amount of power transmitted over the wire. Due to power loss, the load must always be greater than the demand.

3. "Generation" is the amount of power that the plant creates in order to meet that demand.

4. Ways to match generation with load include: anticipate demand, get power from other power plants, have enough reserve power, and perform load shedding.

5. The most common types of reserves are batteries, natural gas, gasoline, water reservoirs, and spinning reserves.

6. Batteries are effective as reserves for small power producers, but are not effective for major utility companies.

7. Natural gas can be stored for long periods of time, and can be converted into electricity in a relatively short time.

8. Reservoirs for hydropower systems are effective reserve systems, such as by opening additional sluice gates to provide more power, or using small hydropower sites only for peak use.

9. Facilities can use closed-loop hydropower systems where one type of power production (such as wind) pumps water into the reservoir as well as providing power to the community. The hydropower can then be used as needed.

10. Diesel is effective as a back-up source of power if the primary source fails. Diesel can be stored forever, provides a significant amount of power per volume, and can be converted to electrical power very quickly.

11. Spinning reserve is extra power created by the generators at the power plant in addition to the scheduled power. Spinning reserve is available immediately. However, any unused spinning reserve is a wasted use of energy.

12. In order to ensure quality control, all utility and grid managers must monitor the factors of current, voltage, temperature, and frequency. Managers must also be able to operate equipment remotely, and communicate with workers anywhere in the region.

13. When a power line fails we can switch the current to go through an alternate line. The switching is done through relays.

14. It is not uncommon for power plants to shut down temporarily. This is much like circuit breakers which are installed in the home to shut power off rather than damage the equipment.

15. Restarting a power plant takes time; it is not as simple as turning on a switch. It is orderly, but it does take time.

16. We must have a source of power to restart the power plant. There are several ways we can get this power: Diesel generators, batteries, pneumatic devices, house units, and other working plants on the grid.

10.2
Monitoring and Communications Systems

Introduction

In order to maintain reliable electricity, a utility company must have a good monitoring and communication system. In this chapter we will examine the details of monitoring electricity.

List of topics for this chapter

1. Basic Components of a Monitoring and Communication System
2. SCADA Overview
3. SCADA Operations: Control Center Perspective
4. SCADA Operations: Electronics and Wires Perspective
5. Translation Devices
6. Options for Communication Connections
7. Broadband Systems
8. Back-up Systems for Communication

Basic Components of a Monitoring and Communication System

In general, a quality monitoring and communication system for utilities or grids will have the following:

1. Devices which convert phenomena such as voltage, frequency, and temperature into a signal.

2. Wires which carry the signal from remote locations to the control room.

3. Wires going the other way, signaling back to the devices and telling the devices what to do.

4. Phone communication with all utility workers in the field, and phone communication with other power plants.

5. Backup communication methods, for both machines and workers, in case any one line of communication fails.

6. Backup power, to supply power to the communications, when there is no power coming to the normal communication system.

SCADA Overview

SCADA is essentially the high-tech, computerized monitoring system of the utilities. Specifically, SCADA provides information about voltage, frequency, and temperature from various locations and sends that information to a central computer in the control room. The term "SCADA" is an abbreviation for Supervisory Control And Data Acquisition. The entire remote monitoring and remote communications system for any one utility is known as its "SCADA system."

There are two ways to look at the SCADA system: 1) control center perspective, and 2) electronics perspective. We will discuss the details of common SCADA systems below.

SCADA Operations: Control Center Perspective

The SCADA system has a control unit which is linked with various remote units. The control unit "asks" each remote unit to report any changes. The time it takes to ask each remote unit and get information back is very quick, only a second or two is needed to communicate with all remotes.

If the operator needs to respond to changes happening in the electrical system, then he follows a procedure. This set of steps is much like your computer asking "are you sure?" before the final act. This set of double checks exists so that no errors are made.

In many utility companies these actions are automatically recorded on a hard drive. As a further item of quality control, these changes are often printed out.

Furthermore, in the best SCADA systems, each Remote Unit will automatically report to the control room if any variable (frequency, voltage, or temperature) reaches a certain upper or lower value.

SCADA Operations: Electronics and Wires Perspective

The sequence of steps in the monitoring system of electrical power is essentially a series of translations:

Step 1, Language 1
Measurement is taken (volts, frequency, or temperature).

Step 2, Language 2
Measurement is translated into another physical form, often as pulses. (This translation is done using a transducer).

Step 3, Language 3
Measurement is converted to digital form (using an analog-to-digital converter).

Step 4, Language 3
Measurement is sent over the wires, usually as pulses in a binary code.

Step 5, Language 3
Computer reads digital message.

Step 6, Language 4
Computer turns measurement into English on the screen.

Step 7, Language 2
Digital message is also turned into analog, for lights and gauges (using a digital-to-analog converter).

Translation Devices

Introduction

Working with machines means that we must be able to understand the language of machines. The main translation devices that we use in order to communicate with machines are: 1) transducers, 2) analog-to-digital converters, and 3) digital-to-analog converters.

Transducer

The transducer is basically a translator. It translates one mechanical language to another. For example, the telephone uses a transducer. The transducer in a telephone converts sound into electric current. There are numerous types of transducers.

In the arena of utilities the quantity being measured is turned into one of the following: voltage, frequency, or pulses. As a specific example, if the temperature of the power line is 50 degrees Celsius, the transducer might convert that value (i.e., translate that value) to be "5 volts."

Note that if the desired quantity is in the same form as the translation, then the "translation" appears as a smaller, proportional value. For example 70,000 volts might be "translated" in the transducer to 70 volts.

Analog vs. Digital

It is important to understand the difference between analog and digital. Anything referred to as "analog" is a physical factor. Anything referred to as "digital" uses a code.

Analog methods were the first to be developed. In the distribution of electricity, the most common analog signals are voltage and frequency.

Digital methods were developed later. The digital signal uses a binary code, which is a code that is easily understood by computers. There are many ways to create a digital code, but in electrical systems the digital code is usually created by a series of pulses.

Digital is more accurate and more efficient than analog. All new equipment uses digital, and most of the older equipment is gradually being replaced by digital equipment. However analog devices still exist and many work effectively.

Analog-to-Digital Converters

Remember that digital systems and analog systems are essentially two different languages. A digital device will not understand an analog signal. Therefore if we wish to communicate with a computer, then the language of analog must be translated into the language of digital.

In order to turn an analog signal into a digital signal you need an analog-to-digital converter. One typical converter begins with an analog signal in the form of pulses. These pulses create a code, in particular a binary code. The binary code is a digital language which computers readily understand.

Digital-to-Analog Converters

Although computers are effective for processing information, sometimes it is effective to have additional cues for humans, such as lights and sounds. These lights and sounds are usually produced by analog methods. However, as stated above, digital and analog are essentially two different languages. An analog device will not understand a digital signal.

Furthermore, any instructions sent from the control room to the remote equipment begins as digital, yet must communicate with the device as analog.

Therefore, in order for the message from a digital device to be understood by an analog device, we must install a digital to analog converter. The digital-to-analog converter works essentially the same as the analog-to-digital converters, just in reverse. The digital signal comes across as a code. The code is then converted into a physical factor, such as voltage or frequency.

The analog (physical) factor can be almost anything, it just depends on what factor the analog device understands most effectively.

Options for Communication Connections

Introduction

In order to get signals from the remote unit to the control room, we need some form of communication connections. Similarly, in order to get signals from the control room to remote units, we need some form of communication connections. There are three basic options: telephone wires, power wires, and microwave dishes.

Telephone Wires

Telephone wires are connected from each remote unit to the control room. In many areas, you will see an additional wire on the poles under the three wires that carry the phases of power. This wire is the telephone wire that connects remote units to the control room. (Note that if you see multiple such lines, these may be standard telephone communication lines. Yet these phone lines may also include utility communication lines).

The telephone wire uses electrical current. Sound is converted into electricity, the electricity travels down the wire, and is converted to sound on the other end. Therefore one advantage to using the telephone wire for SCADA signals is that voice can be carried over the same line as data.

Another advantage to using a telephone wire for communicating with devices is that communication signals are kept separate from the delivery of power. (This contrasts with the use of power wires as discussed below).

Power Wires

It is possible to carry the SCADA communication signals over the same wire that we transmit the power. Both the electrical power and the SCADA signals are electrical current. Therefore, these signals can be combined and transmitted at the same time.

We "superimpose" the electricity of the communication signals over the electricity of the power. This is called modulation. The modulation in utility communication is similar to modulation for radio. In brief, modulation involves combining a small signal with a large signal, like passengers getting on a bus. The "bus" then travels to the destination, where the people get off. In the same way, SCADA signals and the electrical power are combined, delivered to the same destination, then separated for their individual use.

One advantage of transmitting over the power wire is that we have all the wires in place. There are no new wires needed. However, whatever problems there are with the power wire itself may affect the reporting signals trying to come across that wire. Also, the acts of modulation and demodulation require additional technologies. Therefore, it is often preferable to have a separate phone line or a microwave system rather than signaling over the power line.

Microwave and Cellular Transmission Systems

Just as we have satellites and cellular transmission for phones, we can have satellites and cellular transmission for utilities. In this system, messages are sent through the air on streams of electromagnetic energy.

The main advantage to the microwave system is that fewer communication wires need to be put in. We can connect each remote location with the control room simply through a network of dishes.

Another advantage is the use of point to point communication. It is much simpler to contact the particular device directly using the dish communication system or the cellular communication system.

Broadband

In electronics and communications, the term "band" refers to the number of different frequencies that can be carried over one wire at the same time. When a wire can carry large amounts of different frequencies at the same time, it is considered broadband.

Every communication line in the world is designed to carry a certain amount of different frequencies. The number of frequencies that a communication line needs depends on the amount of information that must be sent over that line. For reference: the telephone wires can carry about 3,000 frequencies; and a television channel can carry up to 6 million different frequencies.

For utility communications, it is essential that the communication system is a broadband system. We need to get signals representing measurements of all types, we need to get signals from many remote locations, then we need to send signals back over those same lines to tell the devices what to do. In addition, it is best to have voice capabilities over the same line as the data.

Therefore, for all these reasons, it is important that we have a broadband system. It is important that our communication system carry as many simultaneous frequencies as possible.

The exact number of frequencies which can be carried in a particular broadband system will vary from system to system, depending on the particular technologies used. Furthermore, as engineers create more sophisticated equipment, the number of frequencies in a broadband system will continue to increase.

Whenever a new communication system is being created, the designers should always use the latest technologies (largest broadband system) which allow for the greatest number of simultaneous frequencies. The exact amount of frequencies to be carried over the broadband system should also be discussed by local utility companies and local governments.

Back-up Systems for Communication

Utilities must have back-ups for of all their signaling and communication systems. Most large phone companies have back-up generators. They do this because they know that communication is essential, and much more so in an emergency. In addition to the phone company back-ups, all utility companies must have back-up generators for their own communications. A power company should never rely solely on the generator of the phone company.

Chapter Summary

1. In order to maintain reliable electricity a utility company must have an effective monitoring and communication system. All key factors must be monitored throughout the transmission system, problems must be fixed quickly, and all parties must communicate effectively.

2. Supervisory Control And Data Acquisition (SCADA) is the remote monitoring and signaling system used by utilities.

3. Communication between remote equipment and the control room is done through a series of translations, using a series of transducers.

4. SCADA signals can be carried over three possible systems: telephone wires; power wires; or microwave/dish network.

5. The term "band" in communications refers to the number of frequencies that can be carried at the same time. Due to the complexity of electrical systems in a region it is essential that SCADA communication systems be as broadband as possible.

10.3
Quality Control for Utilities:
Causes, Practical Effects, & Protection

Introduction

When looking at the quality control for electricity, we look at three main factors: voltage, frequency, and temperature. In this chapter we will go over the basic concepts, terms, and practical points of each factor. Furthermore, for each quality control factor we will look at the following: causes, important effects, and methods of protection.

List of topics for this chapter
1. Voltage Variation Basics
2. Voltage too Low or No Voltage – for Short Duration
3. Voltage too Low or No Voltage – for Long Duration
4. Voltage too High – for Short or Long Duration
5. Frequency Variation Overview
6. Frequency: Changes in Frequency
7. Frequency: Additional Frequencies
8. Temperature too High

Voltage Variation Basics

Introduction

There are three basic variations of voltage that can occur: voltage is too high, voltage is too low, or there is no voltage. The specific names we give to the above three variations depend on the length of time they last. For example, if the variation lasts less than 1 minute, it is said to be "short duration." If the variation lasts more than 1 minute, it is said to be "long duration." In addition, there are other specific names to variations in voltage.

The following are the names of three basic voltage variations (too high, too low, or none) for each duration length (short time or long time):

Short Duration (less than 1 minute)

Voltage variation	Name
voltage too high	swell
voltage too low	sag
no voltage	interruption

Long Duration (longer than 1 minute)

Voltage variation	Name
voltage too high	overvoltage
voltage too low	undervoltage
no voltage	sustained interruption

Voltage too Low or No Voltage for Short Duration

Introduction

Here we consider the events when voltage is too low or there is no voltage, but only for less than a minute. In practical terms, think about your light flickering, or the power goes off then back on very quickly.

Note that low voltage and no voltage factors are similar so they are often treated together.

Causes

Temporary low voltages (sags) are usually caused by short circuits. Any equipment beyond the short circuit does not get as much power as normal. Hence, the user experiences a short duration interruption (sag). These short circuits may be caused by many things, but the most common reasons are arcing in circuits or flashovers on overhead lines. Note that if the wires reach a certain lower level of voltage then a circuit breaker automatically disconnects the line from the system.

Practical Effects

The basic practical effect of a period of sag is that the power to your home or business will be low power or intermittent power. How big of a problem this is depends on your use of the power. For home owners this is not much of a problem. Usually you won't even notice. At most, you might see a flickering light, maybe your television will go out for a second, or your digital clock might flash "12:00".

The more significant effects are on businesses, particularly manufacturing. Manufacturing often has machines which are sensitive to even the smallest fluctuations in power. Manufacturing processes are stopped in the middle of operations, and equipment sometimes stops functioning altogether. Therefore, a temporary surge to these companies is not just a second of trouble; it may take hours to restart processes and repair equipment.

Prevention and Protection

These low voltages (sags) are usually temporary and often fix themselves. If not, turning the circuit breaker back on (which reconnects the wire) will usually show that everything is fine. However, often these short circuits need extra help to be fixed. The solution is "fast tripping" where the circuit breaker is flipped off and on again, many times, very quickly. This usually clears the problem in the line, and the power continues as normal.

The most effective protection against sags and temporary interruptions is the Uninterrupted Power Supply. The Uninterrupted Power Supply is a device that can be put onto a machine so that the machine can ride out those times of low voltage. If a manufacturing company thinks it is wise to have an uninterrupted power supply, then they can have this designed into the equipment. Alternately, it can be bought separately from another vendor and added to the system.

Voltage too Low, or No Voltage for Long Duration

Introduction

Commonly known as a blackout, no voltage for long periods of time is the primary concern of most users. Related to this are the times when the power is reaching the device, but the voltage is too low to be effective (the appliances will not function although technically power is delivered).

Causes

The following are the main causes for the voltage being too low or no voltage for long periods of time:

1. Trees falling on lines
2. Insulators weakening
3. Demand exceeds supply

4. Fuses and circuit breakers responding to abnormalities

5. Lightning which strikes on or near electrical equipment

6. Power plant failures

Practical Effects

The problem area is technically called the "fault." If the fault is near the homes, then only a few people will notice. In contrast, if the fault lies close to a substation or at a high transmission line then thousands or millions of people may be affected. Therefore, the area affected depends on where the problem is located.

The duration can be from minutes to hours depending on the specific fault and the resources of the company to fix it.

Finding Problems in Cables

Because cables are underground, it can be difficult to find the location of the problem. Also, as more and more systems are underground, finding the underground faults is a significant factor in quality control for electrical systems. The best modern method for finding the location of problems is the fault indicator. Fault indicators are placed along the power line, usually in each pad mounted transformer (which sit above ground). The indicators are switches that are flipped by a specified amount of current. The operators can compare locations of indicators switched to those not switched in order to better pinpoint the exact location of the problem.

There is one important factor to consider when using fault indicators: the sensitivity of the switches. Different people suggest different sensitivities; each utility must decide the value for itself.

Protection

1. Trim trees

2. Clean insulation

3. Replace insulation after years of use

4. Add shield wires (see protection against lightning below)

5. Install line arrestors (see protection against lightning below)

6. Install thick wires to hold trees

7. Have back-up communication systems

8. Have back-up generators and enough reserve power

Voltage too High for Short or Long Duration

<u>Causes</u>

High voltage surges are caused by two events: lightning and capacitor switching.

1. <u>Lightning</u>

Lightning is a burst of high voltage electricity. It is obvious to most people that a lightning strike to power equipment will cause damage. In addition, lightning does not have to hit a conductor directly to cause jolts to the power system. Lightning can merely hit anywhere near the lines and cause damage throughout the power system.

There are several ways that lightning can hit near a line (not directly at the line) and still cause trouble: 1) Lightning can strike near a power line and induce a current in the wires. 2) Lightning can strike the ground, and the current will travel through the ground to an underground power line, resulting in a power surge or exterior damage. 3) Heat produced from nearby lighting strikes can burn or melt the outer coatings of an underground cable. 4) Of course, lighting can hit the wires directly, which will result in either instant destruction or a traveling power surge.

2. <u>Capacitor Switching</u>

Capacitor switching is a normal process in the daily operation of a utility company. However, sometimes this capacitor switching can cause long-durations of too high of voltage. (See the chapter on Power Factor in the volume on Efficiency for more details.)

<u>Practical Effects</u>

Practical effects of high voltage bursts are mostly in two areas: transformers and power lines. When a transformer receives a high voltage burst the transformer is usually knocked out of service. When power lines receive a high voltage burst, a wave of very high energy is sent down the wire, which causes major damage until the energy dissipates.

When power lines have a surge of high voltage, such as by lightning, this additional energy will travel down the line. This surge of power exists as very high energy electrons, which transfer their energy by moving or bumping down the power line. The net effect is a wave of electrical power, similar to a tsunami water wave traveling over land.

Like the tsunami, the energy surge of the electrons is often powerful enough to destroy anything in the path, including high voltage transmission lines, substations, and home electronics.

Also, because this surge of energy travels down the line (very high energy electrons are bumping down the line, gradually losing energy along the way) this power burst can travel a long distance before dissipating. Consequently, this surge of energy can destroy several pieces of equipment in the path of the power wave before the energy mellows to a reasonable level.

Protection
There are three main methods to protect against high voltage bursts (including lightning). These include: shielding, line arresters, and surge protectors.

1. Shielding
Shielding is an extra wire on the transmission line. This line is grounded. It does not normally carry power. Lightning will hit this wire, and the wire will carry the burst of power to the ground. Shielding is done primarily for high voltage transmission lines and for underground cables, but not as much for local distribution lines.

Shielding is more expensive, and therefore used only where necessary. For local distribution lines, arresters are more commonly used.

There are two common ways to place these shielding wires: a) along top of the tower, and b) next to each of the power lines. (See figures 9.7, 9.8, and 9.9 in the book on Transmission of Electrical Power.)

2. Line Arresters
A line arrester is essentially a variation of the lightning rod. The arrester is a small device, often mounted on poles, and usually placed near transformers. Some transformers have line arresters built directly into them. (Either way, it is very important for all transformers to have some form of lightning protection).

The line arresters take the electricity from the lightning, then drain the surge at various points along the line. This is known as "bleeding" the current. The extra current is then drawn into the ground, not the homes.

Line arresters are particularly important for underground cables. Lightning that hits the ground can cause significant harm to the underground cables if these cables are not protected.

The placement of line arresters is an important science. To protect the power system from the effects of lightning, we must look at specific placements of line arresters. Factors for placement include: distance from equipment, distance from likely locations for lightning strikes, and much more. A very useful layout is to put a line arrestor on every third pole. In addition, every transformer should have a line arrestor built in.

3. Surge Protectors

Homeowners and businesses can prevent damage to electronic equipment by using a surge protector. In theory, this device will take some of the extra power from the lightning burst. However, many products claim to provide more surge protection than they really do.

Frequency Variations: Overview

Electricity is created as alternating current, which travels as a wave. The particular frequency of the wave is determined primarily by the generator. In the United States the standard frequency of the wave is set at 60 Hertz. Thus, the ideal wave in the United States is one pure wave, cycling at 60 Hertz. However, variations in frequency can occur. There are two main variations with the frequency of electricity: changes in frequency and additional frequencies.

Changes in Frequency

Introduction

The wave of electrical energy should travel at 60 Hertz. However, sometimes the frequency travels faster than 60 Hertz. These are usually short bursts, usually lasting less than a second. These bursts are broadly categorized as low frequency, medium frequency, and high frequency.

Term	New Frequency of the burst
low frequency	less than 5,000 Hertz
medium frequency	5,000 Hertz to 500,000 Hertz
high frequency	greater than 500,000 Hertz

Causes

The main causes of frequency changes are problems with the generator. Changes in frequency can also be caused by problems within the power plant, or problems from other power plants on the grid.

Practical Effects

A general principle of frequency usage is the following: different pieces of equipment must have the same frequency of electricity in order to work together. When power plants are on a grid, and they wish to share power, then that power must be on the same frequency. Some power plants can stall, particularly steam power plants such as coal and nuclear power, if the frequency is changed by even a few Hertz.

Protection

It is important that all power plants and all utility companies closely monitor the frequency through the lines. Any variation in frequency should be adjusted immediately. This is done primarily by: 1) adjusting the frequency of the generators at the power plants, and 2) by using specialized equipment at substations which will normalize fluctuations of frequency.

Additional Frequencies

Introduction

Ideally, the power wave should be just one pure wave. However, sometimes other waves occur on top of the desired one. The ideal pure wave is called the fundamental. Additional waves are called harmonics or interharmonics. Harmonics are waves which have frequencies that are exact multiples of the fundamental frequency. Interharmonics are essentially the same as harmonics, except that these are not exact multiples of the fundamental frequency.

Causes

Additional frequencies are caused by circuitry somewhere along the system. The main causes of harmonics are related to the following: impedance, capacitance, reactance, changing speeds of motors or other devices, changes in frequency, or short circuits.

However, the specific causes of harmonics involve areas of circuitry which I feel would not be of use to the reader.

Practical Effects

There are two main areas that can be adversely affected by harmonics: utility power systems and industries.

1. Utility Power Systems

Harmonics will affect power systems, primarily transformers and telephone lines. Regarding transformers, the harmonics add significant heat, which can easily damage the transformer.

Regarding telephone lines, these harmonics can interfere with communication. The ninth harmonic (540 Hertz) and above are the frequency ranges used for voice communication. Therefore, the harmonics from power lines can interfere with the quality of nearby telephone lines.

2. Industries

Earlier we discussed the "Uninterrupted Power Supply". These devices exist in the machines in order to ride the intermittent low voltages. However, these devices have the disadvantage of distorting the power wave by adding frequencies. Therefore, the power which enters the actual manufacturing equipment may have harmonics and interharmonics rather than a pure wave.

Note that solutions to this problem are mostly up to the engineers of the particular industries rather than the arena of utilities.

Protection

There are three basic ways to deal with harmonics:

1. Add filters which suppress or block harmonics
2. Alter the frequency of the system
3. Reduce harmonic currents which are produced within the machines of the user

The processes involve more circuitry than is appropriate for this audience. If you are interested in the details you can read other texts.

Temperature Too High

Introduction

Temperature is a key factor on the limits of electrical lines and equipment. Note that very high temperatures are more serious than very low temperatures. If a power line or a piece of equipment overheats then that power line or equipment will become damaged, permanently destroyed, or cause electrical fires.

As we proceed with our discussions remember that temperature is a measurement of kinetic energy, and that heat is a primary method for transferring that energy.

Causes

There are two main sources of heat energy: internal and external. The total temperature on the line is a combination of the two sources, and this total temperature must not exceed the design limits.

Internal heat energy: Electricity flowing through wires naturally produces heat. There are two ways that this occurs: 1) greater resistance creates more heat energy, and 2) more current creates more heat energy. The amount of current can be changed by the operators. The amount of resistance is fixed by the choice of material in the wire.

External heat energy: The temperature outside the wires adds to the total temperature. When the air is warmer, as in the summer, then that heat energy is added to the system. When the air is cooler, as in the fall or winter, then little or no heat energy is added to the system.

Practical Effects

Each line is rated for certain temperature limit. We must monitor the temperature of the line, and make sure that we do not exceed that temperature. When the temperature of an electrical line or equipment reaches its limit, then the line will become damaged.

The overall temperature of an electrical line is a combination of the internal temperature and the external temperature. This has a practical effect on how much power we can send over the line in the different seasons.

Let us use a simple example. Suppose that our line is rated for 100 degrees. This means that we cannot exceed 100 degrees within our wire at any given day or time.

Consider a summer day: let us say that on a summer day the temperature from the outside makes the wire 80 degrees, without any current inside. This means that we only have 20 degrees of temperature range left. Therefore, the amount of power we can put through that wire must not raise the temperature more than 20 degrees.

Now consider a winter day: let us say that on a winter day the temperature of the outside makes the wire 40 degrees. This means that we have a 60 degree temperature range left. Therefore, we can put in lots of power into our line, just as long as it does not raise the internal temperature of our wire in excess of 60 degrees.

Therefore, we can put more power through lines in fall and winter (when the air is cooler) than we can during the summer (when the air is hotter). Stated another way, we cannot run as much power through the lines in the summer (when the external temperature is higher) as we can in the winter (when the external temperature is lower).

Protection

The main protections against temperature problems are:
1. Designing equipment to resist high temperatures
2. Cooling the equipment
3. Using fuses and circuit breakers

1. Designed to Resist High Temperatures

One of the most important design factors for any line or any piece of equipment is to withstand high temperatures. This is where the selection of materials can make a big difference. Other design factors include: allowing the internal heat to escape, and insulating the equipment from absorbing external heat.

2. Cooling Systems

Related to the issue of temperature and seasons is the issue of temperature and our cooling of the wires. If we cool the lines and equipment then we can get more power or deliver a higher voltage. Many lines and equipment are designed with cooling of the equipment as a primary consideration.

Air is a natural coolant. The circulating air naturally cools overhead wires. However, underground wires do not have this coolant, and so we must put some coolant into the cables. Equipment such as transformers and substations also need coolant.

3. Fuses and Circuit Breakers

Utility operators must continuously monitor the temperature of the equipment and make adjustments. However, automatic systems must be installed as well. Fuses and circuit breakers exist to protect the lines and equipment from damage, by automatically shutting the item down when the temperature in the system becomes too great. All of these automatic systems are usually set to disconnect the problem at 80% of maximum temperature value.

Chapter Summary

1. When looking at the quality control for electricity, we look at these main factors: voltage, frequency, and temperature.

2. There are upper and lower limits, for each of the above three factors, and for all lines and equipment in an electrical system. These factors must be monitored for each piece of equipment and kept within proper limits.

3. Temporary low voltages (sags) or interruption
    a. Causes: short circuits
    b. Protection: "fast tripping", Uninterrupted Power Supply

4. Long duration low voltage or interruption
    a. Causes: trees falling; insulators weakening; demand exceeding supply; circuit breakers responding to abnormalities; and lightning.
    b. Protection: trim trees; clean insulation; replace insulation; use shield wires; install line arrestors; install thick wires to hold trees; install fences to keep animals away; install a reliable communication system; install reliable back-ups and reserve energy.

5. <u>High voltage, short or long duration</u>
   a. Causes: lightning; capacitor switching
   b. Protection: shielding; arresters; surge protectors

6. <u>Frequency: changes in frequency</u>
   a. Causes: problems with the generator; problems from other power plants on the grid.
   b. Protection: adjusting the generators at the power plants; using quality control equipment at each substation.

7. <u>Frequency: additional frequencies</u>
   a. Causes: circuitry such as impedance, capacitance, reactance, changing speeds (of motors or other devices), frequency changes, and/or short circuits.
   b. Protection: Reduce harmonic currents produced by the machines of the user, add filters which suppress or block harmonics, or alter the frequency of the system.

8. <u>Temperature</u>
   a. Causes: internal heat (from resistance and current); external heat
   b. Protection: Design equipment to resist high temperatures; cool lines and equipment; use fuses and circuit breakers.

10.4
Basic Concepts of Grids

Introduction

A grid is the connection of two or more different utility companies. In practical terms, this means that a utility company can share its power with other utility companies on the grid. This also means that customers get portions of their electricity from many power plants, not just one.

There are advantages to a grid, but there are disadvantages as well. The principle of a grid is good. However, the scale and the operation of grids must be completely revised.

List of topics for this chapter
1. Overview of Grids
2. Grid System is Similar to the Roadway System
3. Regulations, Economics, and Grids
4. Advantages and Disadvantages of Grids
5. Ideal Grid at a Glance
6. Independent System Operators and FERC
7. The Players and Their Roles
8. Additional Grid Terms

Overview of Grids

Introduction

A grid is the electrical connection among multiple generating facilities, multiple utility companies, and multiple end users. More specifically, the entire network of generating plants, utilities, power lines, substations, and end users, within one designated region, is called a "grid".

The grid system is very similar to a complex network of roads. Just as cars need roads to travel, electrical current needs power lines to travel. In the same way that cars merge onto freeways and exit at certain points, the electrical current merges onto major power lines and exits at different points. This entire network power lines within a region is the grid.

Furthermore, someone has to manage the network. Just as we have regional and state transportation authorities who manage the network of roads, we have regional grid operators who manage the network of power lines. The grid operators act as neutral parties which provide oversight and mediation regarding every aspect of the grid system.

In practical terms, the grid system means that customers get portions of their electricity from many power plants, not just one. This also means that a complex market of electrical power exists (buying, selling, trading) between power plants, utility companies, and end users.

There are advantages to a grid, but there are disadvantages as well. The principle of a grid is good – it ensures more power on demand, more customer choice, and greater competition. However, the scale and the operation of grids must be completely revised. Our current grid system is vulnerable to terrorist attacks and the system is more likely to have large scale blackouts. Therefore our grid systems need to be completely changed.

National Grid and Grid Regions

Many people refer to a "national grid." People use this term because electricity is so interconnected in the U.S. that we practically do have a national grid. However, in actuality, there are really several different grid regions.

Grids are officially divided into regions by FERC (the Federal Energy Regulatory Commission). Each grid region is operated by an Independent System Operator (ISO). ISOs are independent, third party operators, which operate the grid in their region. The ISOs are regulated primarily by the Federal Energy Regulatory Commission. There are currently 12 ISOs in North America.

The Grid System is Similar to the Roadway System

Introduction

The grid system is very similar to the network of roads and highways. In the roadway system, individual cars begin from various destinations. Many of those cars merge on the same major roads and then onto the major freeways. Later, the cars get off the highway then to the minor roads, finally reaching their ultimate destination.

In a similar way, electricity is generated from several different locations, then merges onto the wires, travels many miles, and exits at various points, until ultimately reaching the final destinations. Therefore, the grid system is very similar to the roadway system.

In addition, there are many other similarities. In particular, both systems have a traffic flow which needs to be managed daily. Also, both systems have infrastructure which needs to be created and maintained regularly.

Power Delivery Process

An overview of the power delivery process is as follows: The electricity is generated at several power plants. This electricity travels down a series of wires, often merging at major substations. From there the power enters major power lines, and travels many miles. Closer to the destination, electricity gets off at different exiting wires, and travels to the final destinations of homes and businesses.

Traffic Control

Now think about this: how do cars merge onto the roads and highways? The cars can't go all at once or there will be chaos. Therefore we have stoplights and other traffic signals which control the flow of cars.

Also consider the following questions: Who decides where to put these stoplights? Who decides on the timing of the lights? The answer is a regional transportation authority. Similarly, who responds when a road is damaged? Who creates a detour while repairs are being made? Again, the answer is a regional transportation authority.

The same is true for power lines. Who creates the rules for traffic flow of the electrical current? Who decides when a bundle of current is allowed to pass through a region of wire? Who is responsible for detours in the grid system? The answer to all those question is the grid operators.

Indeed, traffic control is such an important role that the ISOs use several oversight mechanisms to ensure smooth flow of electrical power. For example, ISO control rooms have walls with exact replicas of the geographic region; these walls display the events on the grid in real-time. In addition, grid managers are in constant communication with the dozens of individual utilities and power plants throughout the vast region served.

Customer Perspective: Ordering the Same Car from Different Cities

Throughout most of this book we look at the grid system from a bird's eye view, or we look at it from the perspective of the grid manager. In addition to these perspectives, it is important that we understand the grid from the customer's perspective.

We will again use the analogy of the roadway system, but in a slightly different way. Imagine you want to buy a Chevy Impala. You can buy this car from many dealers, and from many locations. Same car, same features, and it will be delivered to your door using the same roadway system. However, as a customer you can buy it from many places. Perhaps you like the price better from one location. Or you like the company culture of one dealer. Or the date the car will be delivered is the best for you. Maybe you are just partial to one city over another. For whatever reason, you can choose to buy that same car from a variety of locations.

Similarly, grids today allow for a competitive marketplace. Local utility companies and local cities can choose which power generating company to buy from. This choice is primarily based on price, historic success record of the generating company, or type of power source.

Furthermore, in some areas individual citizens can sign up with their local utility company to choose where they get their power (such as from only renewable sources or companies which are more socially conscious).

All of this is allowing end users more options in this highly competitive marketplace for electrical power.

However, this also means that the grid managers and state regulators have had to adopt a new major role: marketplace manager. The grid managers must ensure a fair marketplace for buying and selling electrical power. In this marketplace, all members of the public will have complete access to information about power production, cost of power from various sources, and power company data. This information allows all players in the electrical power marketplace, including end users, to purchase the power they want from the sources they want, and at the prices they want. Indeed, overseeing a fair market is another very important role of the grid operators.

Who Owns and Operates the Infrastructure

Electricity flows most effectively through wires rather than haphazardly through the air. Therefore wires are absolutely needed. Similarly, cars travel most effectively on paved roads, rather than all-terrain over the hills. Therefore, just as we need a designated roadway system for cars to travel, we need a designated system of power lines for electrical current to travel.

Now consider these questions related to roads: Who creates the roads? Who maintains the roads? Who gets access to the roads? Then consider that these same questions can be asked for electrical power lines: Who paid for the power lines? Who controls access to the power lines? Who is allowed to distribute power over the power lines?

The answers can be quite varied, and the players involved will affect the fundamental nature of the grid system. We will discuss this in detail later, in the section on the possible players and their roles. At this point note that whoever built a piece of the system has authority to allow or deny users of that piece of the system. They also have the authority to charge fees for others to use that piece of the system.

The existing arrangement of who owns what piece, who has authority, and what fees to charge, is a result of several factors. These factors include: private enterprise, government agencies, regulations, legislation, and an intense amount of haggling between all the parties.

Also note that no two regions are the same in this regard. While we will provide some general concepts and possibilities of grids, for the exact details the reader will need to look at his own grid region and his state's Public Utility Commission.

In general, much of the grid system today is like toll roads. One entity built a section power line, others pay to use it, and a third party manages the daily operations (including maintenance and fee collection). There are of course variations, but that is the most common authority structure today for power lines in the grid system.

Regardless of who owns what piece of the grid system, it is important to remember that the grid managers provide final traffic control and quality control oversight over the whole power transmission system in the region.

Regulations, Economics and Grids

Introduction

When discussing grids we must discuss economics, regulations, and other factors which have nothing to do with science. These concepts are all related; we cannot separate one from another. Grid operations are influenced as much by regulations and economics as by technology.

Regulations and Grids

Regulations will often dictate the types of contracts which can exist between power producers and utilities. These regulations will either encourage a power company to expand, or discourage the company into remaining stagnant. Regulations will also dictate the types of generating facilities which are built (such as wind and natural gas, but not coal or nuclear). Therefore many regulations will directly affect the amount of electrical power generated in the region.

Regulations will also often dictate the methods in which power can be bought and sold. These transaction methods can have a significant influence on reliability – this is because the ease and accuracy of any power trading method has a direct impact on the amount of power delivered each day, and therefore has an impact on reliability.

Regulations may also set rules of market oversight. Therefore, many of the ISO's market operations are done in response to those regulations.

Economic Realities and Grids

Economic realities affect every aspect of the grid system. For example, economic realities of the time (such as financial status of the company) will influence any investments in new technologies or new power plants. Consequently, the number of plants and new technologies (or lack thereof) will then affect the future reliability of the grid.

Additionally, the number of facilities will influence the complexity of grid operations, including traffic control, quality control, and market oversight.

Furthermore, electrical power is the center of economic prosperity in many communities. By developing a strong economic base of reliable electrical power the entire community will prosper.

Regulations, Economics, Technologies, and Grids

Indeed, the procedures for grid operations come not just from the science of power transmission, but also the limitations imposed through government regulations and the reality of economics. Therefore, any discussion on the operation of grids must necessarily mention regulations and economics. Throughout the book, you will see how these factors affect the reality of grid systems.

Advantages and Disadvantages of Grids

Advantages of Grids

The primary advantage of a grid is power sharing. When one company has more power than another, then that company can sell the extra power. The other company buys power to help meet its needs. It is in this way that more people in the country have power at any given time.

The other primary advantage of a grid is that a utility can shop around for the cheapest electricity at any given time. Similarly, in some states customers can shop on the grid themselves for the best deal. With such competition, electricity will be as cheap as possible.

Disadvantages of Grids

The primary disadvantage to grids is the enormous disaster that arises from being interconnected. The existing grid system makes large scale blackouts more commonplace. It is analogous to a million swimmers being tied together – when one goes down, he takes the others down with him. The scale of these disasters would not be as large if the power systems were not as intertwined as much as they currently are.

The grid system is also highly vulnerable to terrorism. A small bomb on a few key locations or a cyber-attack in a few control rooms could put half our nation into darkness.

Finally, even when the grid is operating effectively, the large distances which power travels automatically means we waste large amounts of power each day.

Therefore the entire view of grids must be completely revised. By revising the grid system we will make our grids more reliable, as well as grow our local economies. All of the concepts related to the future of grids will be discussed in detail throughout the book.

Ideal Grid: At a Glance

Introduction

Considering the major flaws the grid system today, it is imperative that we revise our entire grid system. The following are the factors that comprise an ideal electrical grid:

1. Local Power: 80%

A minimum of 80% of power delivered to any community should be from local sources. For this discussion, "local" would include anything within the state boundaries, and preferably as close to the community as possible. Purchasing electric power from long distance locations should be avoided.

Reasons include: a) the community will be more self-sufficient, b) the local economy will prosper, and c) blackouts will remain localized.

2. Small Power Plants Should Provide Most of the Local Power

Hundreds of small power plants should comprise the primary power sources, in total, for that community. The technology and costs today make it realistic for many small power plants, together, to provide the electricity needs of any town or city.

Reasons include: a) the community will be more self-sufficient, b) the local economy will prosper, and c) blackouts will remain localized.

3. Build New Plants in Each Local Area

Sometimes states purchase power from another state. This is not acceptable. Any area which wants power cannot expect another state to provide it. Each state will have to share the burden of new power plants.

Reasons include the primary ones: the community will be more self-sufficient, the local economy will prosper, and blackouts will remain localized. In addition, this criteria is important because shorter distances will reduce power loss along the wires.

Furthermore, when a one state creates the power, then the overall financial cost, environmental cost, and regulatory burden is not felt by the users in the second state. In addition, when each state provides its own energy needs all the factors are truly understood by the decision makers and the citizens in that state.

4. Grid Size Should be Limited

Grid size must be limited. If the grid size is limited then the scope of the blackouts will be limited. When one power plant fails or a major power line is damaged, this will affect only a small portion of the population.

Similarly, a physical attack or cyber-attack on the power grid will only affect a small portion of the nation, leaving the vast majority of the nation with full power.

Smaller grids will also be easier to manage. Smaller regions, smaller networks, and fewer customers will all translate into a grid system that is more easily (and more effectively) managed.

Approximate grid limits can be made by state boundaries, geography of power producing areas, or population served. (The details of sizing grids will be discussed later in this book).

Further Details

The reasons behind each of these Ideal Grid criteria will be discussed in greater detail throughout the book, culminating in the section on the "Future of Grids".

Independent System Operators and FERC

Introduction: ISO Overview

Grids are officially divided into regions by FERC (the Federal Energy Regulatory Commission). Each grid region is operated by an Independent System Operator (ISO).

ISOs are independent, third party operators, which operate the grid in their region. The operation of an ISO is a combination of traffic control center, regional stock exchange, and quality control auditor.

The ISOs are regulated primarily by the Federal Energy Regulatory Commission. ISOs are regulated secondarily by each state's Public Utility Commission.

In order for operators to become an ISO, they must apply to FERC and get official approval. Currently there are only 12 regions of ISOs in North America. (See Appendix)

Regional Transmission Organizations (RTOs)

"RTO" is an abbreviation for Regional Transmission Organization. The functions of RTOs and ISOs are essentially the same. Both RTOs and ISOs are designated by FERC. Both RTOs and ISOs are responsible for operating the grid in their region. The "RTO" is a new type of grid organization, with a few additional authorities (primarily related to the oversight of the power market). FERC is encouraging all ISOs to apply to become RTOs. Many people refer to both RTOs and ISOs simply as "ISOs."

Jobs of the ISO

The operation of an ISO is a combination of traffic control center, regional stock exchange, and quality control auditor. Specifically, the ISO has the following primary jobs:
1. Coordinate the transmission of power throughout the region
2. Supervise the buying and selling of electricity
3. Ensure that there is enough power to meet all demand in the region
4. Manage unexpected events so that customers experience minimal power loss

A good ISO will also do the following:
1. Be fair in buying and selling electricity
2. Welcome small suppliers
3. Welcome environmentally friendly power plants
4. Provide information to the citizens as necessary

Complexity of the ISO's jobs

As stated above, the operation of an ISO is a combination of traffic control center, regional stock exchange, and quality control auditor. Any one of these tasks is complex. All three tasks together can make the daily management of the ISO very complicated.

Furthermore, the complexities of the ISO's jobs are in proportion to the size of the grid. The jobs of the ISO are also time sensitive (response time is often essential within the hour, or within minutes!)

Hence, again, we return to the theme of this unit: grids are good, but smaller grids are better. The complexity of the jobs of any grid, the fast response times needed, multiplied by the miles of transmission wires and numbers of power plants, makes these super larger grids very difficult to manage. We must scale down the size of the grids so that we can manage each grid more effectively.

The Players and Their Roles

Introduction

The arrangement of the grid depends on the players involved. The following are the possible owners for power plants, power lines, and other pieces of equipment.

To begin our discussion, know that in any grid there are essentially three types of entities involved:

1. Suppliers (power plants)
2. Local Utility Companies (power lines and retail selling of power)
3. ISO or RTO (grid managers)

Supplier (Power Plants)

The suppliers are the entities which actually generate the electricity. Suppliers are the companies which own the coal power plants, the hydroelectric plants, the nuclear plants, and the wind farms. These suppliers generate power and sell that power to local utilities.

Each power supplier makes agreements with the local utilities, which helps all parties do business more effectively. These agreements usually specify:
   • How much electricity to produce at any given time, on any given date
   • How much money they will get for their electricity

The possible types of ownership for power generation are:
1. Large private company, usually financed by investors
2. Municipal (city owned)
3. Federally owned
4. Co-op
5. Small private companies
6. Individual home owners and landowners

Owners of Transmission Lines

Transmission lines are high voltage wires which carry electrical power over many miles. Distribution lines are lower voltage wires which carry power throughout a city or in regional neighborhoods. Who owns and maintains these power lines? The possible types of ownership for power lines are:

1. Large private company, usually financed by investors
2. Municipal (city owned)
3. Federal owned
4. Co-op

Operations Manager for Transmission Lines (ISOs)

While any one of several entities (listed above) may own a segment of high voltage transmission line, the daily operations of that power line usually fall to the Grid Managers.

Remember that most transmission lines are required to carry power from a variety of sources. Also note that many of those transmission lines were financed by state or federal agencies. This means that power from a variety of companies will be sent through those wires. Consequently, the only entity that can manage the traffic flow of the transmission lines must be a government approved entity.

Hence the grid manager (the ISO) provides oversight of daily operations for all transmission lines in the region. For example, when one transmission line goes down, the ISO knows about it, and sends the power on an alternate route.

Note that the actual repair of power lines is usually done by another party, such as a local utility company, a local government entity, or other contracted service. ISO makes the call, then the contractor makes the repairs.

Merging Power onto the Transmission Line

Remember that just as with cars merging on the freeway, so it is with electrical power coming from a variety of places that must merge onto the high voltage power lines. This merging takes place at substations.

As stated in the book *Transmission of Electrical Power*, the substation has many purposes. The primary purpose of the first substation is to combine the power from multiple sources into one single power line.

We begin our discussion with the original scenario in the history of electrical power, where one power company owns multiple generating facilities as well as the power lines. Often this one company will have its own substation, where power from multiple generating facilities arrive. The power is combined (like merging cars on a freeway), and then sent down a series of wires to the customers.

Today the situation is similar, yet more complex. Today many power suppliers exist rather than just one, and the power from all those companies must be merged onto the same power line. However, this cannot be done by an individual company, because that company could treat some suppliers more favorably than others. This job of merging oversight necessarily falls to an independent third party: the grid manager (ISO) and its substations.

Therefore, most of the power generated in the region is combined at substations, which are owned and operated by the ISO. This is where the physical operations of traffic control take place.

The power is then sent down the high voltage transmission lines (which are also managed daily be the ISO). This power is then sent to each of the local utility companies (see below), who in turn distribute the power to local customers.

Local Utility Companies

A local utility company is the entity which actually delivers power to customers. The local utility company buys power from suppliers (though the ISO marketplace) then delivers the power through local distribution lines. The local utility company usually owns the local power lines, and is responsible for daily operation and maintenance of those lines.

Many states have reduced monopolies by mandating that if the utility company owns distribution lines as well as power plants, then the power distribution service must be operated as a separate business from the power generation. The two operations must be treated as separate entities in the management and accounting.

In any case, the local utility company must buy their power and schedule power delivery from the power plants (including from "their own generators as a separate business"). These transactions are supervised through the ISO.

Note that some people use the term "utility company" for either distribution service companies or generation companies. However, for clarification in discussions it is best to use the term "utility" only when referring to the distribution service company, and the term "supplier" when referring to the generating company.

In addition, note that a "power company" can refer to either the distribution service company or to the power generation company.

ISO or RTO

The ISO is the entity which controls the grid. The ISO is the grid manager. The ISO provides general oversight and coordination as needed for all activities related to the transmission of electrical power along the grid.

As stated earlier, the ISO performs several important functions, including: traffic control, marketplace management, and quality control. Each of these tasks is important, and some tasks (such as traffic control and market oversight) can be quite complex. Therefore we will describe the details of these grid operations in later sections.

Additional Grid Terms

There are a few additional grid terms which citizens and legislators should know. These are the most common terms found in reports and books regarding the structure of grids which have not been discussed previously.

Additional Grid Terms A: Physical Structure

Capacity

The term "capacity" is used in two different areas: power plants, and transmission lines. When referring to power plants, the capacity is the maximum amount of power the plant is capable of producing, under ideal conditions. When referring to transmission lines, the capacity is the maximum amount of power the transmission line is capable of carrying, under ideal conditions.

Distributed Generation

The term "distributed generation" refers to small, locally owned power plants. Whenever we discussed small, locally owned power plants in this book we were in fact discussing distributed generation. The term distributed generation is used mostly by the Department of Energy. The term is also used by some industry analysts.

Distributed generation can refer to many types of power plants, including:

- small power plants used by a local utility to serve a local region

- small power generation units built on-site at the same location where the power is to be used (such as at manufacturing facilities, at universities, and at hospitals)

- small power plants owned by a local community which serves only that local community.

Islanding

Sometimes a utility company will take power off the grid. This is called "islanding."

Additional Grid Terms B: Agencies and Utilities

IOU – Investor Owned Utility

Investor Owned Utilities are companies owned by stockholders. These private companies are funded by investors, who in turn get stocks and dividends (financial return on their investments). These are usually the largest of utilities. An IOU can refer to either the local utility company (the company which owns the transmission lines) or to a company which owns power plants.

PUC – Public Utility Commission

Each state has a Public Utility Commission (PUC). Most PUCs are independent government agencies. This means that the Commissions are not subject to review by the state legislature. Roles vary from state to state, but common roles of a state's PUC include the following:

a. Approval of electric power plants and electric power transmission lines.

b. Supervising construction and maintenance of power plants and electric power transmission lines.

c. Setting the retail price of utilities. This means that the PUC ensures that rates are reasonable, yet at the same time the rates allow the utilities to earn a fair profit.

Note that in recent years some states have deregulated the price. If the price is deregulated, then the final price is determined through a complex system, supervised by the ISO. Also note that if the state has deregulated the retail price of electricity, the state will still regulate reliability and safety.

QF – Qualifying Facility

A Qualifying Facility is a small power producer which meets special requirements in order to be allowed to distribute electricity along power lines owned by other companies. The concept of the Qualifying Facility comes from PURPA, The Public Utility Regulatory Policies Act. Facilities must apply to FERC and be approved by FERC to become "qualified."

Grid Terms C: Economics and Grids

Market

In general, a "market" is an opportunity to sell a specific product. A "power market" is the opportunity to buy and sell electrical power. Most power markets are supervised by the ISO. Note that an ISO may run several types of power markets at the same time. These markets include weekly, daily, and hourly.

Power Exchange (PX)

A Power Exchange is any place where electric power is bought and sold. Typically, a Power Exchange is part of the regional ISO. The Power Exchange is usually the part of the ISO which focuses on the financial transactions of electricity (vs. the parts of the ISO which manages the scheduling or quality control.)

Spot Market

A "spot market" is any power trading done in less than a 24 hour period. There are typically several categories of markets each of which are spot markets. These include: a) day-ahead, b) hour-ahead, and c) ten minutes ahead.

Note that the contrasting method of power sales to the spot market is the use of contracts. (Any trading done in periods greater than 24 hours is done through long-term and short-term contracts).

Chapter Summary

1. A grid is the connection of two or more different utility companies. In practical terms this means a grid is the connection of several power plants which are owned by different companies.

2. Grids usually connect power plants which exist in different regions.

3. Due to the nature of the grid, the electricity which reaches a customer comes from dozens of utility companies. The total electricity which a customer gets is usually an accumulation of several portions from several different power plants in several different regions.

4. In many grids the power that a customer receives comes from plants hundreds of miles away.

5. Grids exist in order to buy and sell power. There are two advantages:
   a. Providing enough electricity for each community
   b. Shopping for the cheapest electricity at any given time

6. The primary disadvantage to grids is the enormous disaster that arises from being interconnected. When one plant goes down, the other plants on the grid also go down.

7. Grids are good, but they must be done on a smaller scale. The interconnections must also be arranged differently.

8. The arrangement of the grid depends on the players involved.

9. The possible types of ownership for power generation include: large private company, municipal (city owned), federally owned, co-op, small private companies, individual home owners, and landowners.

10. The possible types of ownership for power lines include: large private company, municipal (city owned), federally owned, and co-op.

11. An ideal grid will have:
    a. 80% of power from local or state sources
    b. Much of the power comes from small power plants
    c. New plants always being built in the local area in order to meet demand
    d. Grid size limited by geography or population served

12. Grids are operated by Independent System Operators (ISOs). The ISOs are under the authority of the Federal Energy Regulatory Commission (FERC). ISOs have the following primary tasks:
    a. Coordinate the transmission of power throughout the region
    b. Supervise the buying and selling of electricity
    c. Ensure that there is enough power to meet all demand in the region
    d. Manage unexpected events
    e. Be fair in buying and selling electricity
    f. Welcome small suppliers
    g. Welcome environmentally friendly power plants
    h. Provide information to the citizens as necessary

13. The "capacity" is the maximum amount of power the generating facility is capable of producing, or the maximum amount of power the transmission line is capable of carrying.

14. The term "distributed generation" refers to small, locally owned power plants, including:
- small power plants which serve a local region
- small generation units at manufacturing facilities, universities, and hospitals.
- small power plants which serve local neighborhoods and small communities.

15. When a utility company takes its power off the grid this is referred to as "islanding".

16. Investor Owned Utilities (IOUs) are private companies funded by investors. These are usually the largest utilities. An IOU can own power plants or transmission lines.

17. Public Utility Commissions (PUCs) are independent government agencies. Roles vary from state to state but typically include:
- a. Approval of power plants and transmission lines.
- b. Supervising construction and maintenance of power plants and transmission lines.
- c. Setting the retail price of utilities.

18. A "Qualifying Facility" (QF) is a small power producer which meets special requirements under PURPA in order to be allowed to distribute electricity along power lines owned by other companies. Facilities must be approved by FERC to become "qualified."

19. A "power market" is the opportunity to buy and sell electrical power. Most power markets are supervised by the ISO and may include separate markets for weekly, daily, and hourly scheduling.

20. A "Power Exchange" (PX) is any place where electric power is bought and sold. The Power Exchange is usually the division of the ISO which focuses on the financial transactions of electricity (in contrast to the divisions of the ISO which manage the scheduling or quality control).

21. A "spot market" is any power trading done in less than a 24 hour period. There are typically several categories of spot markets, including day-ahead and hour-ahead.

22. Long-term contracts are the preferred method for buying and selling power. Most of the trading done in periods greater than 24 hours is done through long-term and short-term contracts.

10.5
Grid Operations

Introduction

Grids play several major roles in the transmission of electrical power. These roles include traffic control, market oversight, and quality control. In this chapter we will discuss the specific operations of grids in greater detail.

List of topics for this chapter
1. Basic Grid Operations
2. Variations of Grid Operations
3. Ancillary Services

Basic Grid Operations

Grid operations can be very complex. We will begin our discussions by looking at the simplest version of grid operations. The following is the basic sequence of operations in a power grid:

1. The Local Utility Estimates How Much Power it Will Need

This estimation is usually determined 24 hours in advance. The estimation is detailed, specifying how much power to be delivered, by specific times (often listed in 1-hour blocks), on a specific date.

2. The Local Utility Finds a Supplier

The utility company goes shopping to find a supplier (the supplier is a company which owns and operate power plants) which will provide that electrical power. The local utility usually chooses its suppliers with any of the following criteria: price, ability to meet the power needs, and history (track record of reliability).

3. The Utility Company and the Supplier Agree to Terms of the Contract
   The utility and the supplier agree on specifics, including:
   • How much power to be delivered at any specific time, throughout a specific 24 hour date.
   • The price to be paid for that electricity.
   • The specifics of how the transaction of money will take place.

   The agreement can be for any period of time. Most companies prefer a 1-year contract, although seasonal contracts and monthly contracts are also used.

4. This Purchase Agreement is Sent to the ISO
   The contract is sent to the ISO for evaluation and processing. The ISO is a neutral third party, but there are some items for the ISO to consider, including reliability standards and scheduling. The ISO may also need to evaluate the contract for adherence to PUC regulations. The ISO may also acts as the banker.

5. The ISO Considers the Schedule, and Does Traffic Control
   Usually there is a "Scheduling Coordinator" who manages this task. The Scheduling Coordinator puts the information into a sophisticated computer program. This program compares the proposed schedule to other schedules already approved. If there are no conflicts, then the schedule is accepted. However, if the transmission lines are already being used by other companies at that time, then the proposed schedule must be adjusted.

6. Supplier Creates Power from the Power Plants
   Once the schedule is accepted, the supplier knows that it can go ahead and start creating that power.

7. The ISO Watches the Flow of Power Throughout the Grid
   This is done from a centralized control room. If problems occur, then the operators at the ISO will take necessary action. These actions can include changing the path of the power, requesting more energy from a local supplier, making calls to local utility companies, or any number of other actions.

8. The Financial Transactions Take Place

As with many other service industries, the utility company usually pays the power supplier after the power has been delivered.

There are several methods for the money transfer to occur. For example, some ISOs have a separate financial division which acts similar to the PayPal system: the ISO receives money from one company, and sends the money to another company, charging a modest handling fee in between.

Some companies set up standard billing between them, which leaves the ISO out of the transaction. However, the ISO will often step in as financial manager for specific types of transactions.

The ISO may also act as bill collector or neutral third party in a financial disagreement.

The specific method for payment usually depends on:

- the ISO's Operating Procedures

- the agreement between the local utility company and the supplier.

Variations of Grid Operations

Introduction

Every step described above is what occurs in the simplest situation. This is the ideal, but the reality is that there are numerous variations. Furthermore, there are many factors which complicate the procedures very quickly. In this section we will look at a few of the ways in which the grid operations can vary from the simple scenario above.

Long-term Contracts and Advance Scheduling

Most utility companies like to sign contracts with various suppliers. Everyone benefits from these contracts:

- Local utility companies know that they can count on a certain amount of power, every day, for at least a year.

- Suppliers know that they have a steady buyer for their product, every day, for a year.

• Electricity is generally cheaper, because suppliers will offer discounts for utilities which sign long-term contracts.

• Suppliers can plan more effectively for the production of power.

• Daily schedules are already made.

• The ISO has an easier job of coordinating schedules.

• Suppliers are more willing to build new power plants, because the steady income (guaranteed in contracts) gives the suppliers the cash flow they can count on.

Therefore long-term contracts are preferred by most companies in the energy industry.

Utilities Sometimes Under-Estimate Power Needs; Suppliers Plan for More

Local utilities often underestimate the demand for power in their territory at a particular time. Therefore, the local utilities need to get that power from somewhere, and get it fast. This is usually not a problem because many suppliers are prepared for this situation. Many power suppliers will make more power than they originally sold and scheduled just for this potential situation.

Let us look at both sides from the beginning and watch the process. Utility companies figure what they need, and schedule that amount of power from the power plant. Suppliers (power plants) create that power and deliver it at the time promised. Yet, in addition to the scheduled power, many power plants create additional power. (See "Ancillary Services" below.) Many power plants figure that somewhere in the network a utility will have underestimated their needs.

Indeed, most of the time a utility company somewhere has underestimated the power needed for a particular time. Suppliers have planned for this, and offer that additional power for sale. Additional power is made, then purchased, very quickly. This power is sent along to the local utility very quickly. Thus, the supplier will be making utilities happy and make customers happy by avoiding blackouts.

At the same time, the supplier will be making a nice profit. Because this power is needed quickly and the buyer has an immediate need, the seller is in a nice position and the cost is much higher than normal.

Power Bought by ISO

Often the ISO will own a certain amount of power. This is done to prevent large scale outages.

In this situation, the ISO acts like a local utility company. The ISO is always getting a certain amount of power from various large power plants. This is a type of spinning reserve, which is used for emergencies. If a major problem occurs along the lines, the ISO has the option of sending their power (the ISO's power) to those utilities. This is done so that additional power can be sent very quickly.

Note that if any of this power is sent to a utility which needs it, then that utility will be billed for that power, at the current market price.

Ancillary Services (Reserves for Sale)

Introduction

Ancillary services are power reserves available for sale. These reserves are always available in very short times, usually between a few seconds and a few hours. These types of power reserves are always bought during the day the power will be used. There are many types of ancillary services (reserves for sale during the day). The specific types of ancillary services and the specific terms for those ancillary services depend on the particular ISO.

Ancillary Services (reserves) vs. Surplus Power

Ancillary services (reserves) are specifically created as such. Note that an ancillary service is not the same as surplus power.

Surplus power is when the plant creates the power the local utilities need, but it turns out that one of the utilities, for part of the day, does not need all the power that the utility requested. This power can be sold by the power supplier (since it has already been made) but it is not considered an "ancillary service." In contrast, ancillary services are specifically created to be sold on the ancillary services day-of-use market.

Summary

1. The basic grid operation is done in the following steps:

   a. The local utility estimates how much power it will need.

   b. The local utility then goes shopping to find a supplier (a power plant) which will provide that electricity.

   c. The local utility company and the supplier agree on specifics such as the amount of power to be delivered at specific times and the price.

   d. This agreement is then sent to the ISO.

   e. The ISO first considers the schedule. If there are no conflicts with other transfers already scheduled on the transmission lines, then the schedule is accepted.

   f. Once the schedule is accepted, the supplier knows that it can go ahead and start creating that power.

   g. Then the financial transactions take place. The specific methods depend on the operating procedures of the ISO.

   h. The ISO watches the flow of power throughout the grid from a centralized control room. If problems occur, then the operators at the ISO will take necessary action.

2. Actual grid operation is much more complex. Some of the complicating factors include:
   a. Utilities under-estimate power needs
   b. Unexpected situations occur
   c. There are several markets, at different times.
   d. There are thousands of power related companies on one grid
   e. Grids often serve millions of people
   f. Grids often supervise thousands of miles of transmission lines
   g. Grids often cover several states
   h. Weather varies day to day, year to year, and region to region

3. Most utility companies like to sign contracts with various suppliers. Long term contracts benefit power plants, ISOs, utilities, and customers.

4. Local utilities often under-estimate the demand for power in their territory at a particular time. Therefore, the local utilities need to get that additional power from somewhere, and get it delivered quickly.

5. The ISO will often buy a certain amount of reserve power. This is done in order to prevent large scale outages.

6. Ancillary services are power reserves available for sale within 24 hours of use. An ancillary service can be delivered quickly, usually between a few seconds and a few hours.

7. The operation of the grid can become vastly more complex in relation to the size of the grid.

10.6

Quality Control for Grids

Introduction

Due to the size of most grids one unexpected event could cause power failures over a large area. Therefore, it is essential that a grid has effective quality control.

<u>List of topics for this chapter</u>
1. Quality Control of Grids
2. Grid Failures
3. Need for More Power Plants and Smaller Grids
4. Sabotage and Terrorism
5. NERC and Electric Reliability Regions
6. FERC and NERC: a Brief Comparison
7. Energy Policy Act of 2005

Quality Control of Grids

<u>Introduction</u>

One of the primary functions of the ISO is to monitor the electricity throughout the region. When unexpected events occur, such as power line damage or power plant failures, the ISO must respond.

In general, quality control for a grid is the same as the quality control for a utility. Therefore, almost everything discussed previously on quality control for utilities will apply to the quality control of grids. In the next sections we will see how these principles apply specifically to maintaining quality control of grids.

1. <u>Match Power Generation with Demand</u>

One of the primary roles of an ISO is to see the big picture of power use throughout the region. Therefore, the ISO helps all the power companies to match generation with demand.

The ISO helps to ensure that power generation meets demand primarily by estimating how much power the region will need for the day, and reporting that estimate on the ISO website.

The power plants throughout the region read these estimated values on the ISO website. The power plants can then use this data to help estimate how much power to produce.

ISOs predictions are similar to weather predictions, and in fact demand for power is closely related to the local weather. ISO first gives broad estimates for the season and month-in-advance. These estimates come from weather predictions from NOAA, and by looking at past data for power use at the same time of year.

The ISO then provides weekly and 3-day advance estimates of power demand. These estimates are based on weather predictions from NOAA, as well as from major events such as festivals.

Final estimates are posted 24 hours prior to day of use. These estimates are the most accurate.

Remember that some amount of time is required to increase power production. Although the generator creates power quickly, the furnace or reactor requires some time to increase production. Therefore having accurate estimations, days ahead of intended use, is very important.

2. Create Enough Reserve

The ISO has oversight of creating enough reserve power. The ISO fulfills this role by encouraging the trade of ancillary services.

Recall from earlier that ancillary services are power reserves available for sale, always in very short times, usually between a few seconds and a few hours. These types of power reserves are always bought during the day the power will be used.

Also recall from earlier, that in any ISO the ancillary services marketplace includes many types of reserves for sale during the day. The specific types of ancillary services, and the specific contractual terms for those ancillary services, depend on the particular ISO. In total, this ancillary marketplace allows any utility company to purchase additional reserve power which they may need on that same day.

In addition to utilities purchasing ancillary services (the day-of-use reserves), many ISOs buy their own power reserves from the power plants. The ISO acts as its own utility company, buying additional power from power plants when needed in order to meet regional demand.

3. Build Quality Electrical Lines and Distribution Systems

Many ISOs are building and maintaining their own transmission lines. In these cases the ISO is 100% responsible for building power lines which are very efficient and are capable of withstanding extreme weather.

In addition, many transmission lines are built and maintained by the individual utility companies, not by the ISO. However, all transmission lines are part of the grid, therefore the ISO must keep an eye on the quality of transmission lines being constructed. Some ISOs have official oversight, while others have a less formal arrangement.

4. Know the Limitations of Lines and Equipment

Exceeding the specified limits of current, voltage, power, or temperature will permanently damage power lines and other equipment. Therefore, the ISO must know the limits of power lines and equipment on the grid (particularly transmission lines and underground cables) and ensure that these lines never exceed the specified limits.

There are thousands of miles of power lines in any ISO, and these lines are of varying voltage and capacity. Therefore the ISO generally prioritizes the power lines they monitor, in this order:

1. High Voltage Transmission Lines: built by the ISO or a government entity.

2. High Voltage Transmission Lines: built by private companies

3. Substations and Relays: for High Voltage Transmission Lines

The remaining power lines and equipment are usually monitored solely by the local utility companies.

5. Have an Advanced Monitoring and Communications System

Each utility company has its own advanced monitoring and communications system. In addition, each ISO has a similar sophisticated monitoring system.

Key factors such as current, voltage, power, temperature, and frequency are measured at numerous locations throughout the grid. This data is measured at transformers, substations, and segments of power lines every few seconds. This information is then sent back to the ISO control center.

In the control room, individuals monitor all the quality control factors for specific regions of the grid. Each individual has his own computer to monitor his area of responsibility.

In addition, the ISO has a large map of the region, with images of transmission lines and power flow on the walls. The control room will either have large video screens, or lights on a wall map, or both. These screens and wall maps show major power lines, flow of power, major substations, and key relays throughout the grid. This visual system allows everyone to see the large picture of power flow, and to see the location of any major problem.

Many ISOs have their own communication lines. This means that ISOs and utility companies are getting individual communications regarding the same region of power line or same substation at the same time. Such information allows the ISO an additional tool for quality control.

6. Switch Transmission Paths Quickly when Any Piece Fails

An ISO installs many relays, alternate transmission paths, and remote communication systems. In the control room the grid operators are able to see on the map where there is a failure. The operators can then switch the relays from the control room to the new path. However, note that electricity travels quickly, therefore the operators at this ISO might not be able to switch relays quickly enough to prevent power failures.

7. Disconnect and Repair Equipment Quickly

When there is a failure, the equipment must be disconnected. If the failed equipment is between the power plant and the utility company's substation, then the ISO will likely be responsible for disconnecting the equipment.

If the failed equipment is between the utility substation and the customers, then the local utility will likely be responsible for disconnecting.

Repairing equipment is usually the responsibility of the entity which owns that equipment. In most cases, the entity which owns the equipment is the local utility company, and the utility company will make the repairs. If the transmission line or substation is owned by the ISO, then the ISO will repair the equipment.

8. Restart a Power Plant Without Difficulty

This is the responsibility of the power plant, not the grid managers. However, the ISO can help by arranging start-up power to be delivered to the failed power plant.

9. Interconnect with Other Power Systems Effectively

The entire structure of the grid should be designed so that power systems can interconnect effectively. There are several areas where the individual power systems should "interconnect effectively":

a. Physical Structure

The primary physical structure for connecting effectively includes transmission lines, substations, unused lines as alternate routes, and relays.

This physical structure also includes various power lines which allow all approved energy companies access to the main line (like an on-ramp to a freeway). In conjunction with the "on-ramp", the grid structure requires a "toll gate" system, operated by the ISO, which dictates whose power gets onto the main line, and at which times. (Note that both of these actions are usually done at substations, and often these particular substations are owned by the regional ISO).

b. Market Structure

The market structure for connecting effectively should encompass a wide variety of methods for energy companies to buy and sell power in a seamless fashion.

In order to ensure quality control of the grid from a market perspective, the ISO needs to create a system where 1) all players have equal access, 2) all data is accurate, and 3) the actual trading of power is as simple as possible.

c. Interactive Websites

The power systems must also be interconnected through interactive websites. The ISO websites must provide accurate estimates for power demand, and accurate background information of power producers.

The ISO must post the planned schedule for power delivery (who, where, and when), and real time viewing of power flow throughout the grid.

Power producers must provide websites, accessible to all players on the grid system, to see how much power the power plants are producing at any time, and the cost of that power.

Utility companies must provide websites, accessible to the ISO, in order to view the utility company's real-time power demand and their monitoring of quality control along key points.

Having such an interactive set of web sites among the various players in the grid system will provide a much more effective dynamic, and ultimately provide better quality control over the entire grid system.

Grid Failures

Introduction

Large grids lead to greater disasters. In this section we will understand why this is so. In brief: electricity naturally flows from where electricity is generated to where there is none. Change the circumstances, and you might reverse the flow. When this situation exists on a grid it can create a cascading effect and create a major blackout.

Sequence of a Large Scale Blackout

1. Power Plant Failure

A large scale blackout begins when a major power plant fails. This power plant stops producing power, and cannot deliver any power to the customers which rely on that power plant.

2. Direction of Flow Changes

The electrical current will flow different directions than normal. Just as nature abhors a vacuum, and water flows through an empty pipe, the electrical current will flow through the wires from high voltage to low voltage. Current flows toward the location where no electrons are being actively pushed.

Similarly, the higher voltage electrons are pushing on the lower voltage electrons, thereby also influencing the direction of any current in the system.

3. Connected on a Grid

If the power plant were isolated then the plant failure would affect just the nearby city. However, most power plants are connected to a grid. Therefore, a power plant failure will affect everyone on the grid system. The main reason is this current will interact with current from other power systems in such a manner as to damage equipment and to change the routes of all current flowing in the system.

4. Smashing Current Creates Voltage Spikes

Furthermore, current is still coming in from some power systems (where the power plants and equipment are still operational) while the original current is flowing backward, which then causes permanent damage to the equipment.

Think of this in terms of cars on the road. Consider two roads merging onto a main highway, with all cars traveling at 70 mph. Suddenly, one group of cars decides to turn around and go the other way...straight into the oncoming cars...with all cars going 70 mph. This results in large scale collisions with serious damage.

In a similarly way, the electrical current going backwards will smash into oncoming current (the current traveling from a working power plant). This creates an excessive surge in voltage, and causes damage to the equipment.

5. Damaged Equipment, Blocked Paths, More Voltage Spikes

Now you have a power line or piece of equipment which is unable to carry electrical power. The current must then travel somewhere else. This is very similar to cars finding an alternate route when the normal route is blocked.

Yet you know what happens when a road is closed: more cars are traveling on the other roads, all packed in. There is potential for more accidents – especially as some drivers are in a hurry. Any such additional accidents can then block additional routes.

In a similar way, the electrical current starts packing up. Then, because the current can't flow through, the energy is transferred through the system as an increase in voltage. This voltage increases until the equipment can't handle it anymore, then BOOM! That second power line or second piece of equipment is now damaged beyond use.

6. Cascading Effect of Alternate Routes, Voltage Spikes, and Damaged Equipment

In this way, current flows where it can, sometimes taking alternate routes, sometimes crashing into current from other paths, sometimes building up and compacting.

The energy from all this smashing and compacting creates a large increase in voltage. These large spikes in voltage will then destroy numerous lines, transformers, and relays throughout the network. The damaged lines and equipment then contribute to further power loss, more forced alternate routes, further increases in voltage, and damage to additional equipment. And then the process continues.

At this point we have a cascading effect: a series of voltage spikes, damaged equipment, and other chaos which brings power failure in multiple regions.

Additional Contributing Factors to Escalating the Blackouts

7. Additional Energy Suppliers Push the Limits

Sometimes the supplier company tries to push the power plant beyond its normal capability. (The power supplier usually does this in order to meet the new demand).

However, power plants cannot be pushed beyond their limits. When this occurs, many interior components will become damaged. (Damage comes from spikes in voltage, excess heat, and/or extensive shaking). This results in destroyed equipment at the plant, and hence another power plant failure.

With no power produced from the plant, there will be large scale power outages for users in the region (even if all power lines are working). In addition, this creates an additional power outage sequence as beginning with #1, which contributes to the problem even more.

8. Substations Drawing Power

Substations can also escalate the problem. When electrical current from reverse flow smashes with current from other systems at the substation, these collisions may cause large enough voltage spikes to damage the equipment. With this substation out of use, no power can be sent beyond that point.

9. Restarting Power Plants on a Grid

There are special considerations when restarting a power plant that is normally connected to a grid. When several plants are down, then several plants must be started up again. However, because these plants are interconnected, and because the users are still create the same amount of demand for power, the process of restarting these plants must be coordinated. If not done properly, it is very possible to repeat the cascading effect, and create the blackouts all over again.

Need for More Power Plants and Smaller Grids

New Power Plants will Prevent Black-Outs

In each of the most recent large scale black-outs, one of the primary contributing factors was the lack of new power plants. Therefore, it is important that every region of the country build enough power plants to meet the needs of its own population. We cannot get electricity by wishful thinking, nor should we put the burden of our needs onto other states.

Back-up Power Plants

A quality utility company will have back-up plants to ensure that customers have power. However, if there are not any back-up power plants, then the power from that producer is shut down to all customers who rely on that power plant. Therefore we must have sufficient back-up generation facilities in order to provide power during any contingency.

Importance of Smaller Grids

It is clear that when just a few power plants fail, and they are connected to a large grid, then the result can be disastrous for millions of people on a large grid. Therefore grids are good, but smaller grids are better. With a smaller grid, the blackout will be more localized.

Sabotage and Terrorism

Introduction

So far in our discussion, we have looked at technical causes for grid failures. However, what about grid failures caused by sabotage or terrorism?

Large Grids Make Us Vulnerable to Terrorism

Large grids create the possibility of large scale disasters caused by terrorism. As early as 1982, the Federal Government had looked into the problem. At that time, the government had concluded that sabotage of just a few key power plants or sabotage of a few key substations could put a great majority of the country in blackouts.

During the last 25 years, the situation has made us that much more vulnerable. There are two main additions to the situation:

1. The grids have increased in size over that time. Many more power plants and many more customers are interconnected than was the case 20 years ago. This means that a small terrorist attack or cyber-security attack could produce a huge result. Under the existing network, a small terrorist attack in key locations will cause power failure for the majority of the nation.

2. Terrorism is more of a genuine danger than in previous years. The terrorism activities of September 11, 2001, has shown that a terrorist attack anywhere in the United States is very possible.

Small Grids are Essential to Our Security

We must have smaller grids, much smaller than we currently have, if we are to be safe from the real threats of terrorism. The very nature of a smaller grid will make an attack on a grid less appealing to the terrorists, and therefore less likely to occur.

Furthermore, if our grids are small and our communities are electrically self-reliant, then a few attacks on utilities here and there will not have any significant effect on our nation as a whole.

NERC and Electric Reliability Regions

Introduction

Another organization which has an important role in quality control of electricity is the North American Electrical Reliability Council (NERC). NERC exists for the sole purpose of ensuring reliable electricity throughout North America. NERC is a voluntary organization, not a government agency. NERC is a non-profit organization.

Members of NERC

NERC members include just about anyone involved with power. Members of NERC include: utilities, grid operators (ISOs), government agencies, and customers. As the name implies, utilities from all three nations (United States, Canada, and Mexico) are involved.

In addition, there are several committees of technical experts. These experts come from all areas of the electrical industry, and from various regions of North America. The total number of experts within NERC is approximately 2,700.

Regional Reliability Councils

NERC has divided North America into Ten Regional Reliability Councils. These Regional Reliability Councils perform the jobs of NERC in the local regions. Each Regional Reliability Council has a say in the development of reliability standards. More important, each Regional Reliability Council assesses utilities and grids in the region regarding reliability and following NERC standards. Each Regional Reliability Council can inspect and enforce violations in reliability.

How NERC Achieves Reliability in North America

NERC achieves its goal of reliable electricity primarily through:
- Creating reliability standards
- Education of those standards
- Voluntary compliance of those standards
- Enforcement of those standards

In total, NERC does all of the following things:
- Sets standards for reliability of electricity
- Monitors, assesses, and enforces compliance of reliability standards
- Offers education and training regarding reliability
- Coordinates activities with Regional Reliability Councils
- Certifies reliability service organizations and personnel
- Coordinates protection of critical grid infrastructure
- Helps make information exchange easier and more effective

FERC and NERC: Brief Comparison

The following is a comparison between the Federal Energy Regulatory Commission (FERC) and the North American Electrical Reliability Council (NERC).

1. FERC is part of the federal government. NERC is a voluntary organization.

2. FERC mandates standards through law. Compliance with NERC is voluntary.

3. FERC enforces standards through fines. NERC also enforces best practices with fines, but can only impose fines on member utilities.

4. Both FERC and NERC encompass North America. (Note that although FERC is a Federal entity, meaning it is legally of United States jurisdiction, FERC makes arrangements with Canadian power companies.)

5. Every location in the United States is within both an ISO region (FERC) and a reliability council region (NERC). Similarly, most utilities belong to both an ISO and a Reliability Council. Also, the local ISO and the local Reliability Council have an intimate working relationship.

Energy Policy Act of 2005

Introduction

A few years ago, Congress passed the Energy Policy Act of 2005. This Act fundamentally changes many aspects of the market structure and wholesale transmission of interstate electrical power.

Electricity Reliability Organization (ERO)

One of the most significant provisions of the law is the power of FERC to enforce reliability. The Act creates a new reliability oversight agency, called the Electricity Reliability Organization (ERO). The ERO will be organized and function very similarly to NERC.

Like NERC, the ERO will divide the United States into regions, operating under regional councils. Like NERC, the ERO will set reliability standards.

However, unlike NERC the reliability standards will be mandatory, not voluntary. Furthermore, unlike NERC, and unlike the earlier authorities granted to FERC, the ERO has official government authority to enforce compliance of reliability standards. Remember that the new Electricity Reliability Organization will be under the jurisdiction of FERC.

The effectiveness of the Electricity Reliability Organization is yet to be evaluated because the ERO is so new. The ERO needs to organize their offices, develop standards, and begin enforcement. The practical operations of the ERO and the effectiveness of that office will need to be evaluated in the coming years.

Oversight of Power Market Manipulation

The Act of 2005 has several other significant elements. One new element is FERC's increased oversight of market manipulation. Market manipulation affects not only price, but through a chain of economic steps also affects a power company's desire to produce power at a particular time or to create new power plants. This, in turn, can greatly reduce reliability.

Although FERC had some authority in these areas prior to 2005, the provisions in this law give FERC substantial increased authority to investigate and impose penalties on those who manipulate the market.

FERC's role in this area is very similar to that of the Security Exchange Commission (SEC). In fact, FERC is using the experience and regulations of the SEC as guidance in this role.

Oversight of Energy Company Mergers and Monopolies

FERC also now has increased oversight over mergers and monopolies. The history of government's involvement in electrical power has been a balance between limiting a natural monopoly and letting the market forces dictate the growth of businesses. Allowing or denying the mergers of power companies will affect not only price but also the ability to produce power, and thus also reliability.

Due to the Act of 2005, FERC has increased roles in making rules regarding mergers, and in enforcement of those rules. In practical terms this means that the players in wholesale electrical power may change. For example, some larger power companies may be required to divest subsidiary companies, particularly if they acquired those companies through unethical practices.

However, many power companies will continue to be able to merge, but only after scrutiny and approval by FERC.

Greater Authority Granting Right-of-Way for Power Lines

Another significant element of the Act of 2005, and a topic significant to this book, is FERC's additional roles in approving transmission lines. The Act of 2005 has given FERC great authorities in granting right-of-way for transmission lines throughout the nation.

The purpose for these authorities is genuine: The construction of new power lines is far below the pace it needs to be to keep up with the demand for electrical power. In some areas, power generation is keeping up with demand, but there are not enough power lines to deliver that power. Lack of transmission infrastructure naturally leads to decreased reliability. Therefore, the United States needs a significant number of new transmission lines to be built.

However, some of the authorities granted to FERC regarding transmission lines are questionable in their Constitutionality. Some of the authorities granted in the law potentially overstep the rights of the States.

Other authorities granted in the law potentially overstep the rights of the people, particularly regarding the abuse of eminent domain. Legal challenges in this area may be expected.

Summary

1. Quality control for a grid is similar to quality control for a utility.

2. The ISO should estimate how much power the region will need for each day and report that amount on their website.

3. The ISO can ensure enough reserve through the ancillary services market. The ISO also ensures enough reserve through buying power reserves from the power plants each day.

4. The ISO should measure key factors such as voltage and frequency, every few seconds, at numerous locations throughout the grid.

5. Relays, alternate transmission paths, and remote communication systems must be installed and maintained along the grid.

6. The ISO can help restart a failed plant by arranging start-up power to be delivered.

7. The entire structure of the grid should be designed so that power systems can interconnect effectively. This includes:
    a. the physical structure of transmission lines and substations
    b. the market structure of buying and selling electricity
    c. the design of the interactive websites
    d. the procedure for scheduling power

8. Large scale power failures occur because electricity naturally flows from where it is generated to where there is none. When no current is being produced you may reverse the flow, which may lead to a cascading effect of alternate routes, voltage spikes, and damaged equipment.

9. Large grids lead to large disasters. In order to prevent large scale blackouts, grids must be smaller and more power plants must be built.

10. Terrorism is a serious threat. Attacks on a few plants or substations could put our nation at great risk. In order to prevent risk from terrorism, our grids must be scaled down to a much smaller size.

11. Quality Control is also effectively ensured by NERC, the North American Electrical Reliability Council.

12. NERC achieves its goal of reliable electricity primarily through:
    a. Creating reliability standards
    b. Education of reliability standards
    c. Voluntary compliance of reliability standards
    d. Enforcement of reliability standards

13. NERC is a voluntary organization. Utilities are not required to join, nor are utilities required to follow the practices. However, most utilities willingly join and want to follow the best practices because no utility company wants a major power outage to occur.

14. Electric Reliability Regions are organized under NERC. Each Reliability Region is supervised by a Regional Reliability Council. Each Regional Reliability Council has the authority to:
    a. Inspect and assess local utilities regarding reliability
    b. Inspect and assess grids regarding reliability
    c. Offer awards for good reliability
    d. Enforce violations in reliability

15. The Energy Policy Act of 2005 has given FERC many new authorities including:
    a. Oversight of market manipulation
    b. Oversight of mergers
    c. Increased authority to acquire right-of-way for transmission lines
    d. Enforcing reliability through the Electricity Reliability Organization

10.7
Smart Grids

Introduction

The term "Smart Grid" is a catch-all phrase for an advanced monitoring and response electrical delivery system. The Smart Grid system monitors overall demand of electricity in a particular region, and then makes necessary adjustments in the power production and the delivery system as necessary.

Because there are many places along the delivery system where the Smart Grid can be used, the application of Smart Grid technology must be evaluated for each area separately. For example, placing individual sensors at private homes must be viewed differently from placing sensors at distribution substations.

List of Topics for this Chapter
1. Overview of the Smart Grid Process
2. Smart Grids as Related to the SCADA System
3. Remote Sensors and Automatic Adjustments
4. The "Smart" in Smart Grid
5. Privacy Issues
6. Improper Use of Smart Grids
7. Proper Use of Smart Grids
8. Voluntary Cooperation with the Smart Grid System

Overview of the Smart Grid Process

There are numerous variations of the Smart Grid. However, the essential elements of most Smart Grids are as follows:

1. Remote measuring devices are placed along numerous locations in the delivery system.

2. Each device automatically registers power demand and sends that information to a central computer.

3. The central computer then automatically makes adjustments in the delivery system to meet that demand.

4. The end result is that everyone gets the power they want, at the hours they want, at a price they are comfortable with.

Smart Grids as Related to the SCADA System

SCADA System Review

Recall from the sections on SCADA (Supervisory Control And Data Acquisition) that SCADA is essentially the high-tech, computerized monitoring system of the utilities.

Specifically, SCADA provides information about voltage, frequency, and temperature from various locations, and sends that information to a computer in the central control room. The operators at the central control room use that information to make adjustments in the delivery system.

In order to make adjustments, the operators send signals to various locations which instruct the device to make a specific change.

The entire remote monitoring and remote communications system for any one utility is known as its "SCADA system."

<u>Smart Grid Similarity to the SCADA System</u>

The Smart Grid system is similar to the SCADA system in some respects. The Smart Grid is similar in that there are sensors placed along the delivery system, and information from those sensors is sent remotely (usually through the air via microwave frequencies) to a central computer. However, a Smart Grid usually differs in the following respects:

1. The Smart Grid measures the *amount* of power being delivered, whereas SCADA measures *quality* of power being delivered.

2. The Smart Grid system is more automatic. Whereas the SCADA system relies on the operator to evaluate the information and make adjustments, the Smart Grid system gives the information directly to a computer, which makes the decisions and makes the adjustments automatically.

3. The Smart Grid is more pervasive. Whereas the SCADA system traditionally places sensors only in a set of key locations, the Smart Grid places sensors in a wide variety of locations – including private homes and businesses. (Note that this brings up some privacy issues, which will be discussed later).

Remote Sensors and Automatic Adjustments

<u>Locations for Remote Sensors and Automatic Adjustments</u>

In the Smart Grid system, remote sensors and automatic adjustment technologies can be placed in many areas. The following locations are the most common:
1. Individual Homes and Businesses
2. Distribution Substations
3. Metropolitan Region Substations (outside the city)
4. Transmission Lines
5. Underground Power Cables
6. Power Plants
7. Power Company Wholesalers and Distributors
8. ISO Regional Offices and ISO/RTO Main Office

Thus, the sensors can be placed at any or all of these locations, sending information about power use to the central computer. In addition, the technologies which make the *actual adjustments* to the delivery system can be placed at any of those points.

Making Adjustments can mean Many Things

In practical terms, what does "make adjustments accordingly" actually mean? Depending on the Smart Grid system, and the location of the adjustment to be made, an "automatic adjustment" could be almost anything. Some of the most common automatic adjustments include:

1. Increasing power production from an operating power plant.
2. Starting a power plant to meet emerging demand.
3. Switching the amount of power received from various sources.
4. Respond to weather to maximize use of solar, wind, and hydro.
5. Switch delivery paths along different power lines.
6. Send more power to one region over another.
7. Reduce the amount of power sent to certain businesses or homes.
8. Disconnect power to specific equipment or outlets.

From this list you can see that the adjustments made could be almost anything. Again, I stress the concept that "Smart Grid" can mean many things. Each technology and each application must be evaluated on its own.

The "Smart" in Smart Grid

The brain behind a Smart Grid is a central computer with a complex set of algorithms (decision-making processes). This computer can be viewed either as an advantage or a disadvantage, depending on how the computer makes its decisions.

One the one hand, a computer can process these millions of signals and run through a series of algorithms faster than any human. Ideally this will lead to more effective power delivery adjustments. On the other hand, taking a human out of the process can lead to unintended consequences. Therefore, regardless of how smart the smart grid is, there will always be a need for humans to oversee the process as a common sense sanity check on the situation.

Privacy Issues

Introduction

There is also a real concern over privacy. There are two categories of privacy concerns: 1) sensors which monitor personal use of electrical power, and 2) automatic adjustments which can disconnect power from an individual's home or from specific appliances.

Sensors as Invasion of Privacy

The purpose of the remote sensor in the Smart Grid system is to measure the use of electrical power at a particular location, then send that information to the central computer. Operators at the control room desire this information to see trends of power use. After being given that information, the operators (and more often the computers) will make adjustments in either the amount of power produced, or the directions which existing power should be sent.

This is the overall benevolently intended use of sensors by well-meaning people. However, the placement of those sensors can intrude on personal privacy.

Sensors can be placed anywhere, including power plants, substations, and transformers. Sensors can be placed along sections of power lines and underground cables. These are generally benign and legitimate placements of sensors.

However, sensors can also be placed outside or inside any business, including within manufacturing equipment. Sensors can be placed in the home, not only outside the home at the meter but inside the home along branches of wiring and within specific appliances. Furthermore, most of these sensors are being installed without citizens being aware (or being allowed to express their opinions).

This may sound fictitious and Orwellian; however I have been involved with enough discussions with well-meaning engineers and policy makers to know that this is true. Furthermore, these well-meaning policy makers continue to push for sensors in more personal use locations.

Therefore, at what point does the sensor intrude on personal space and individual privacy?

Automatic Adjustments

The purpose of automatic adjustments in the Smart Grid system is to make adjustments anywhere along the transmission path, in order to provide power to the users who need it most. Automatic adjustments can include almost anything, as stated in the list above.

Most automatic adjustments are related to opening up transmission paths or changing transmission paths. Many other automatic adjustments are related to acquiring more power from a particular power plant or specific region of the grid.

However, some of those automatic adjustments include disconnecting power from individual users. Power can be disconnected remotely from any home or business. Power can be disconnected remotely from any equipment or appliance.

Although this may sound fictitious and Orwellian, there are many policy-makers who believe in the central planning approach to the grid. This includes disconnecting any power to particular users who "use too much power".

Technologies already exist which allow a central control room to remotely disconnect power from buildings or appliances. These technologies are often installed, either by user choice or in a stealth manner, in many residential and commercial areas. This may be acceptable as a "smart home" where the home owner has complete control over operations (such as starting his dishwasher from a cell phone). However, it is not acceptable for grid managers to take control over those remote adjustment technologies.

Again the question is at what point does the automatic adjustment infringe on personal choice?

Invasion of Privacy through Stealth and Legislation

Grid managers install sensors, install remote adjustment technologies, and acquire control of individual's use of power either through stealth or through regulations.

The stealth mode is generally done by the utility company by coming in and installing the technologies without anyone knowing. If asked, they respond "you voluntary signed up for our service, and therefore you implicitly consented to all of our operations".

The regulation approach involves the grid managers working with the appropriate government agency to develop regulations which mandate that remote sensors and remote adjustment technologies be designed into next generation appliances. This allows the Smart Grid managers to control every individual's use and every business owner's use of electrical power.

This is not fictitious. It is a reality that is growing, without the knowledge of most citizens. Elements such as these in the Smart Grid system may be ethically and legally questionable.

Note that the specific details of the engineering and the regulations vary so much from region to region that the citizens of each community must evaluate the situation for themselves.

Improper Use of Smart Grids

The Smart Grid system can be a significant invasion of privacy and it can lead to an Orwellian State if used improperly. Remember that a proper use of Smart Grids is one that measures overall use at various regions, then adjusts the power supply and delivery options in order to meet the demand at any given time. In contrast, an improper use of a Smart Grid is one that infringes on personal privacy and personal choice in the name of sensible power management.

Improper uses of Smart Grids include the use of smart meters, smart appliances, and automatic control of equipment – especially when those devices are mandatory and automatically controlled at a central level.

I partake in many Smart Grid discussions with professional engineers, policy-makers, and local citizens. I observe many other discussions (where I cannot partake for whatever reason). Too often these discussions focus on technologies which infringe on personal privacy issues. Instead, these discussions should be focusing on the method for monitoring and adjustments on power lines, power plants, and substations.

Well-meaning engineers enjoy placing sensors on private homes, private businesses, specific pieces of equipment, and specific appliances. Many of these well-meaning engineers also talk of automatically shutting off power to specific appliances, specific equipment, and even to specific homes and businesses. I believe that this is a wrong use of the Smart Grid System.

We can summarize the improper use of Smart Grids as follows: An improper use of the Smart Grid is any use which is a personal invasion of privacy.

Any use of the Smart Grid which measures individual use of electrical power, particularly inside a building or at specific appliances, is not a proper use. Any use of the Smart Grid which automatically disconnects power to a user, either to a specific appliance or to the whole facility, is also an improper use.

Note that the Smart Grid managers may use aggregate information (such as at a local substation) to make decisions. The Smart Grid System can also adjust power production and distribution accordingly. However, using technology in the name of "Smart Grid" to monitor and prevent individual use of specific equipment is not acceptable.

Proper Use of Smart Grids

The Smart Grid system can be good if used properly. A proper use of the Smart Grid is one that measures overall use at various regions, such as at regional substations, along transmission lines, and at power plants.

A proper use of the Smart Grid is one that adjusts the amount of power produced and the direction that power will be delivered. A proper use of the Smart Grid may use voluntary cooperation with specific customers to adjust power use (see below). A Smart Grid is used properly when it takes weather into account in order to make the best use of renewable energy production.

Above all, a Smart Grid is used properly when it measures the demand for energy, then makes adjustments in the delivery system to meet those demands at any given time.

Voluntary Cooperation with the Smart Grid System

Another good use of the Smart Grid System is a voluntary partnership with the end user and the Smart Grid. With such a partnership, the end user gets cost savings and more people on the system will have power, easily managed with automatic adjustments.

For example, when there is not enough power produced to meet demand, the end user can have certain pieces of equipment automatically shut down, or some regions of the facility shut down.

This allows the grid managers to have a customer who has volunteered to be given less power when there is not enough power available for the region. In turn, all other customers in the grid will have enough power for their normal power needs. This method of the Smart Grid System will prevent power outages at peak use times.

Everyone will benefit from the opt-in approach, and yet there will be no infringement on individual choice. 1) The general population will benefit because all other people in the system will have enough power for their individual needs. 2) The opt-in customer will benefit from this voluntary cooperation by paying a discounted rate for all power he uses (when power is actually delivered to his facility). 3) The grid managers benefit by having a simple automated system to manage the flow of power during peak use.

Similarly, if the price gets above a certain amount, the end user can have certain pieces of equipment or specific regions of the facility automatically shut down. Again, this will reduce overall demand and therefore allow enough power to be delivered to the remaining customers. The opt-in customer benefits in this situation by paying lower cost overall (never paying above a certain price, and zero cost when no power is delivered). And, as in the case above, there will be more power available to the general population.

Conversely, the opt-in user can set his devices to run only when the price of power is below a certain amount. This is usually when fewest customers are on the grid, and therefore again everyone who pulls power from the system will get enough power for his needs. At the same time, the user will pay the lowest cost by operating some devices at the lowest cost times of day. Furthermore, the times of lowest cost may vary; and by using the opt-in system for lowest cost, the user will not need to program his equipment, he can simply leave the decision to the central control room of the smart grid.

Remember the key is voluntary opt-in partnerships. The Smart Grid must never reduce power or shut off power to customers without a specific opt-in consent. Most importantly, the Smart Grid managers must never shut off power to equipment or facilities automatically based on the decisions of a smart grid system...not without human oversight or the opt-in consent.

Summary of Smart Grids

1. The term "Smart Grid" is a broad concept which includes a variety of technologies. The key elements are:
   a. Remote Sensing
   b. Decision-Making by a Central Computer
          and
   c. Automatic Adjustments

2. Smart Grid sensors and automatic adjustment technologies can be placed anywhere along the system, including specific power plants, substations, transmission lines, relays, buildings, and appliances.

3. The proper use of a Smart Grid is to measure power demand in each region then adjust power production and delivery accordingly.

4. The Smart Grid can also take into account regional weather conditions in order to make the most of renewable energy production.

5. Improper use of a Smart Grid is any use which infringes on personal privacy or personal choice. Mandatory shut-offs and monitoring of personal power use are not acceptable activities of the Smart Grid. However, voluntary partnerships and opt-in arrangements are acceptable.

6. There are numerous Smart Grid technologies. There are numerous locations for sensors and for automatic adjustment technologies. Therefore each Smart Grid application must be evaluated on its own terms.

10.8

The Future of Electrical Distribution: Smaller Grids and Local Power Plants

Introduction

Distribution of electricity will be dramatically transformed in the next decades. Specifically, the future of electrical distribution will be in two areas:

1. Local, smaller grids
2. Local, smaller power plants

List of topics for this chapter
1. A New Direction
2. Benefits of Smaller Grids
3. Sizing of Grids
4. Benefits of Smaller, Locally Operated Power Plants
5. The Need for Some Large Power Plants
6. PURPA: Good for Competition & Environmentalism

A New Direction

Contrary to popular thinking, the "national grid" is the wrong way to go. Popular discussion, including among many policy makers, is that we must have a more complex, more inter-related set of power networks in a vast national grid. This is in fact the exact opposite of the direction we must take.

As we have shown, the trend toward a national grid invites terrorist attacks and is susceptible to power failures on an enormous scale. The "national grid" is a design that will result in large scale catastrophes.

In contrast, the better plan is to create networks that are smaller and regionally focused. Specifically, and as stated above, this requires two directions:

1. Local, Smaller Grids
2. Local, Smaller Power Plants

Note that both of these new directions will require a major change in thinking. Power companies, legislators, regulators, and citizens must begin to think about electrical distribution in new ways. Everyone must think local. Everyone must think smaller. This is the new plan for energy in our lives.

Note also the concept of Smart Grids can be used in this new direction. A Smart Grid is simply a system of remote sensing and automatic adjustments. Such technologies can be used on the future grid systems of any scale.

In our discussion of the Future of Electrical Distribution, we will focus on three areas:
   a. The Benefits of Smaller Grids
   b. Sizing of Grids
   c. The Benefits of Smaller, Locally Operated Power Plants

The Benefits of Smaller Grids

Introduction

For the past century, the common view regarding grids was that larger grids were more effective than smaller grids. However, technology and circumstances have changed. Today, smaller grids are more effective than large grids. In the following sections, we will look at some of the benefits of smaller grids.

1. Blackouts will be Localized

Should a plant failure occur, the blackout would be localized. If the grid size is limited then the failure will affect only those nearby, and not cause the cascading failures (as would be common to a larger grid).

2. Restarting Plants after a Blackout will be Simpler

We discussed earlier how restarting power plants must be done carefully, particularly when connected to a grid. We must be careful how we restart plants when connected to a grid, so that we do not recreate the cascading failures. However, if the grid is smaller then there are fewer plants to coordinate with. The logistics of restarting all plants is easier with a smaller grid.

3. Not Vulnerable to Terrorism

Terrorism is a serious issue. The more interconnected our power plants are, the more vulnerable we will be. Just a few attacks on a few key plants or substations could wipe out much of the power for the entire nation. In order to prevent this, we must go to smaller grids. Using smaller grids, if there were an act of terrorism on a power plant or substation, only the nearby cities would go down. All the other major cities of the nation would be on their own grid, and therefore most of the nation would remain with full power.

Sizing of Grids

Introduction

The most important decision for citizens and for legislators to make is the size of the grid. Each community must decide how many power systems to connect with. If the total amount of power generation is too small, then there won't be enough power to meet the needs of the community. On the other hand, if the power generation and distribution is connected to hundreds of power plants across several states then a large scale disaster is certain to come.

The right size grid will be large enough to provide sufficient power for the community, but will not be connected to so many power systems that large scale disasters can occur. Approximate grid limits can be made by state boundaries, geography of power producing areas, or population served.

Grid Size Limited by State Boundaries

Limiting the grid by state boundaries has many advantages. The two most important advantages to limiting grid size by state boundaries are: 1) regulations are much simpler, and 2) economics are much simpler.

Electrical power in each state is regulated by the state's Public Utility Commission. If a grid crosses several states, then regulations and prices must be negotiated among the Public Utility Commissions of all the states. However, if the grid is contained within one state then the management of the regulations and the prices is much simpler.

Similarly, changes in the economic structure of the grid are easier to do with a smaller grid. Note that there are many options related to the economic structure for utilities. Figuring out all these options is difficult enough for one state, but it is nearly impossible to come to a consensus on the details if the grid covers several states.

Grid Size Limited by Geography

Limiting the grid region by geographic boundaries is a useful approach for small states or less populated states.

Grids can be limited geographically by primary power sources. For example grids can be limited by hydropower regions, wind power regions, or solar power regions. Grid boundaries can also be designated by major geographical barriers, such as mountains or rivers.

Grid Size Limited by Population

If the grid is not limited in any other way, then the grid should be limited by population. A good rule of thumb is to limit the grid size to no more than 10 million residents.

Statistical engineers have determined that a network of power lines, substations, transformers, and power plants which serve 8-10 million people is the maximum size for optimum grid operations. This is the maximum size for optimum response times to unexpected events and for effective oversight of power marketing.

The Benefits of Smaller
Locally Operated Power Plants

Introduction

The second trend of electrical distribution will be in the category of small, locally operated power plants. The benefits for this situation are numerous. Top benefits include:

1. New power plants will be easier to build
2. There will be fewer blackouts
3. More production of environmentally friendly electricity
4. Fewer power lines needed
5. Less power loss, resulting in greater efficiency
6. Local businesses will prosper
7. Local economies will thrive
8. Local communities will be more self-reliant

1. New Smaller Power Plants will be Easier to Build

Smaller power plants can be built more quickly than larger power plants. Therefore by focusing on the construction of small power plants we are more likely to meet the growing demand for electrical power. There are several reasons for this:

a. Smaller Scale Results in Lower Cost

Smaller size power plants require lower cost, and therefore will be easier to build. The amount of material required, skilled manpower, transportation equipment, and specially designed components will be less, resulting in lower overall cost. A lower cost to the company means that the companies can afford to build more power plants in a shorter time.

b. Loans are Easier to Obtain

Banks and investors are more willing to put up a smaller amount of money for smaller scale plants than to put up very large sums for a larger plant. With financing more readily available, energy companies are more optimistic about building new power plants.

c. Smaller Size Plants are Built Faster

Smaller size power plant can be built more quickly, simply due to the scale of the design.

d. Permission is Easier due to Smaller Size

Regulators study the details of power plants before approving. If a plant is larger and more complex, then naturally the time to study, ask questions, compare to existing regulations, and approve will take longer. In contrast, a smaller plant is easier for regulators to study, and the approval process will be much faster.

Similarly, if a power plant is going to be built on a large area of land, or will impact a large segment of the ecosystem, then regulators will be cautious before obtaining approval. In contrast, if a power plant will take only a small area of land, or impact only a small section of the ecosystem, then regulators are more willing to approve.

Therefore small power plants are always easier to approve than large power plants.

2. More Power Plants Result in Fewer Blackouts

Smaller plants can be built more quickly than larger plants. The final benefit will be enough power plants to provide sufficient power to meet the needs of the growing population at any given time.

Remember that the major cause for large scale blackouts (city-wide or state-wide blackouts) is not having enough power plants to provide the power when needed. Therefore large scale blackouts will become less common as we increase the construction of smaller power plants.

3. More Production of Environmentally Friendly Electricity

Smaller power plants are generally more environmentally friendly. Some examples include:

a. Solar Power at Buildings Where Used

Solar panels can be placed on the roofs of (or on the land next to): homes, schools, shopping centers, and manufacturing facilities. The power produced from this solar power will therefore provide electricity to the users in the nearby building.

b. Wind Turbines Provide Power for Local Communities

Wind turbines can be placed on local farms, regional parks, and in specially zoned areas of the county. The power produced from the wind turbines will provide electricity to residents living nearby.

c. Micro-Hydro Systems on Remote Rivers

Micro-hydro power systems can be built along any river, including remote locations, if those rivers have a decent flow rate much of the year. Therefore these micro-hydro systems can provide sufficient electrical power for individual users, small businesses, park facilities, and mountain resorts which lay along these rivers.

d. Burning Biomass and Trash at Many Facilities

Burning biomass and trash can be done within specially designed facilities on many properties, including the properties of ranches, farms, manufacturing plants, and zoned regions of the county. The power produced from this burning will supply some of the electrical needs for the nearby residents, and at the same time will reduce the amount of waste to be managed.

4. Fewer Power Lines Needed and Shorter Wires Used

Local power plants will deliver power to local communities. This means fewer power lines are required.

This eliminates the need for long distance wires. Instead, the power is sent a short distance (usually a few miles, sometimes within a few blocks) from power plant to user.

Having fewer power lines will provide several benefits in our delivery system. Some of the initial benefits include the reduced right-of-way required and the reduced instillation costs. Ongoing benefits include fewer possibilities for lines to become damaged and less maintenance. Perhaps most important, shorter power lines result in greater efficiency (see below).

5. Less Power Loss and Greater Efficiency

As discussed in the book *Transmission of Electrical Power*, every power line will lose power due to inherent resistance in the wire. Furthermore, longer power lines will always lose more power than shorter power lines. Therefore if we want to reduce power loss during the transmission of electrical power, one of the best methods is to use a shorter power line.

Because local power plants deliver power to local communities, the power lines are short – within a few miles, or a few blocks – which is much shorter than the existing transmission lines which travel hundreds of miles. Therefore, the infrastructure of local power plants sending the power over shorter wires will lose far less power (and be far more efficient) than the infrastructure of the long transmission lines presently being used.

6. Local Businesses and Individuals will Prosper

Everyone needs electricity. Small energy businesses can provide this need, and become successful in the process.

a. Smaller Plants can be Built and Operated Anywhere

Unlike large power plants which require large amounts of land and other special requirements, small power plants can be placed almost anywhere. This opens up a vast opportunity for small power plants to be built in any region of the world, regardless of geography, climate, or population. This also leads to the next benefit...millions of individuals will prosper.

b. Many Individuals will Prosper

Smaller power plants will allow more people to prosper, and in multiple ways. For example, local power producers can generate and sell their power to local residents. In this process the owners of the energy company will earn good income and live a quality life. These energy entrepreneurs can achieve all of this regardless of location.

Furthermore, the power provided to the local residents will allow those users to prosper. This power can be used to run local businesses and small manufacturing facilities, which in turn will allow these business owners to prosper as well.

7. Local Economies will Thrive

Perhaps the best benefit of locally produced electrical power is that local economies will thrive. Here are a few points which emphasize how locally owned power plants will benefit the local economy:

a. Self-Sustaining and Interdependent Local Economy

As stated above, entrepreneurs can create energy companies in the community. Their businesses will thrive. Also as stated above, the businesses which use the electrical power can create more goods and services, which will allow those businesses to grow.

In addition, there are all the support functions for the energy company, which are provided by secondary companies. These secondary companies will now prosper. In turn, the secondary companies buy products and services from a third set of companies, which allows those companies to prosper...and so on. This is how a local economy grows and becomes self-sustaining.

b. Base of the Energy Economy Will Remain Steady

Above all, notice the most important feature of this thriving economy: it will never go away. The base of this economy is the production of local energy. This base will never be depleted, and it can never be outsourced.

Unlike other industries like mining, fishing, or agriculture, the base of this economy will never disappear. Wind, solar, water, biomass, and trash provide the energy sources. These sources will last essentially forever.

Also, this base cannot be outsourced. The power is produced by local weather (wind, sun, water) and therefore cannot physically be outsourced. Furthermore, this power is sent through regional power lines to local residents, and again can never be outsourced. Because the base of this economy is physically connected to the community, this economic base will never be outsourced.

Therefore, locally produced and locally distributed renewable energy will provide an economic base which will remain in place for the foreseeable future. The energy company will provide not only electrical power to the community, but economic power to the community as well.

8. Local Communities will be More Self-Reliant

Locally produced power will also allow communities to be more self-reliant. At this time, too many communities depend on their electricity from power plants which are located far away. Therefore, when there is a power plant failure or when there is a failure in the grid system, that failure will affect numerous cities. Residents will be affected who live hundreds of miles away from the original event.

In contrast, when local power plants provide the power for a community, then that community will be more self-reliant. Power plant failures and grid failures will not affect any community which is independent of those systems. The community will be able to stand on its own, regardless of events elsewhere.

The Need for Some Large Power Plants

It must be remembered that for many areas at least one large power plant will need to be built. Smaller plants are often based on renewable energy sources such as the sun or the wind. However, solar power and wind power are unpredictable and there may be times that these will provide no power or insufficient power. Therefore, at least one large plant should be built in each major metropolitan area in order to provide enough electricity on demand, any day of the year. These power plants will likely be coal, natural gas, nuclear, or large scale hydropower.

However, we can keep to our main two trends for better energy networks: smaller grids and smaller power plants.

The ideal scenario is as follows: The renewable energy sources will provide much of the power in the region. This power will be produced and distributed locally.

This power from renewable sources will be supplemented as needed by larger power plants. However, these large plants are a) secondary sources used for supplementing power, and b) primarily supplying power to local communities in the region, with very little power being sent outside the regional area.

In this way we provide all the power we need, yet we benefit from locally produced power and smaller grid systems.

PURPA

Introduction

For almost a century, large utility companies had a monopoly. The utilities built the power lines and the utilities maintained the lines. Therefore these utility companies had no reason to let any other producer of electricity sell their product over those lines.

Smaller companies couldn't get their electricity to the people. Soon these smaller companies disappeared, and there was no business reason to create one. Cheaper electricity and more environmentally friendly electricity were possible, but this electricity could not be delivered.

All that changed with the passing of PURPA. The U.S. Congress passed the Public Utility Regulatory Policies Act (PURPA) in 1978, and it has changed the world of electrical power distribution dramatically. The main provisions of PURPA are:

PURPA Regarding Distribution Companies

1. All utility companies who own power lines *must buy* power from smaller companies.

2. All utility companies who own power lines *must distribute* the power from smaller companies on the power line.

3. All utility companies who own power lines *must pay a fair price* for power from smaller companies. This price is the same as if the larger company produced that same power itself.

PURPA Regarding Smaller Power Plants

1. In order to be able to use the power lines which are built by other companies, these smaller companies must use 75% renewable sources (such as wind, solar, or hydropower) or cogeneration for their total electrical power generation.

2. These smaller companies must also qualify. This means that these smaller companies must be held to a standard of reliability and quality.

3. These smaller companies are not under the same state price regulations as the larger utilities. However, the small companies negotiate their fair price with the larger utility, which in turn (in most cases) is price regulated by the state.

Results of PURPA

Since 1978, PURPA has opened up the doors for several practical results:

1. The net result of PURPA is that smaller companies can distribute their electrical power and can compete in the power marketplace. This has allowed a financial incentive for entrepreneurs to create small energy companies.

2. In turn, this creates greater competition in the energy marketplace, which results in lower cost for the consumer.

3. PURPA also fosters the production of environmentally friendly electrical power. PURPA mandates that the 75% of the power used by these smaller companies is generated from renewable sources. In addition, smaller energy companies are often inherently based on renewable energy. These two factors have contributed to the overall boost of electrical power from wind, solar, and hydro being produced and distributed throughout the nation since 1978.

4. Many small energy companies have been created, many of these smaller companies have grown, and the many individuals who created these companies have prospered due to PURPA. Many more businesses and individuals will undoubtedly prosper in the future.

Chapter Summary

1. The future of electrical distribution lies in two areas:
    a. Local, smaller grids
    b. Local, smaller power plants

2. The benefits of smaller grids include:
    a. Blackouts will be localized
    b. Restarting plants after a blackout will not be as complex
    c. We will not be as vulnerable to terrorism

3. One of the most important decisions for citizens and for legislators to make in electrical power is the size of the grid.

   The ideal grid size will be large enough to provide sufficient power for the community, but not be connected to so many power systems that large scale disasters can occur.

4. Approximate grid limits can be made by:
    a. state boundaries
    b. geography of power producing areas
    c. population served

5. As we reduce the size of each grid, it is essential that we have an emergency line connecting grids. These emergency lines will act as jumper cables for power plants, in case of a natural disaster or an act of terrorism.

6. The benefits of smaller, local power plants include:
    a. New power plants will be easier to build
    b. Fewer blackouts
    c. More production of environmentally friendly electricity
    d. Fewer power lines needed
    e. Less power loss and greater efficiency
    f. Local businesses will prosper
    g. Local economies will thrive
    h. Local communities will be more self-reliant

7. Some large power plants will always be needed. This is because small power plants are often based on wind or solar which is intermittent.

8. PURPA ensures that the distribution of electricity is not a complete monopoly. The main provisions of PURPA are:

   a. All utility companies must buy power from smaller companies.

   b. All utility companies must distribute power from smaller companies on the power line.

   c. All utility companies must pay a fair price for power from smaller companies.

   d. The smaller companies must use 75% renewable energy for their total electrical power generation.

   e. These smaller companies must qualify, which means being held to standards of reliability.

   f. These smaller companies are not under the state price regulations as the larger utilities. However, the small companies negotiate their fair price with the larger utility, which in turn (in most cases) is price regulated by the state.

9. Practical Results of PURPA include:

   a. Smaller companies can distribute their electrical power and can compete in the power marketplace.

   b. Greater competition in the energy marketplace and lower cost for the consumer.

   c. Production of environmentally friendly electrical power.

   d. Business growth and individual prosperity for many people who work in the small energy and renewable energy businesses.

Conclusion

Many Americans hold passionate views about electrical power, yet few Americans understand all the details behind their passion. Electricity should not be mysterious. The science, the technology, and the data of electrical power can be understood by anyone.

Above all else, we must remember that there are no perfect solutions, there are only choices. Any option can be beneficial, yet each option has its own technical issues to work with. It is up to you and to your community to make those educated decisions. I hope that this book will help guide you in your choices.

M.F.

Appendix

1. ISOs/RTOs
2. Regional Reliability Councils

ISOs/RTOs

· Listing is approximately West to East, North to South
· Canadian-only ISOs are listed after ISOs of United States

1. RTO West/Grid West   www.rtowest.com
2. California ISO   www.caiso.com
3. West Connect   www.westconnectrto.com
4. MISO RTO (Midwest ISO) www.midwestiso.org
5. SPP (South West Power Pool) www.spp.org
6. ERCOT ISO (Electric Reliability Council of Texas) www.ercot.com
7. New England RTO / ISO New England www.iso-ne.com
8. NY ISO www.nyiso.com
9. PJM Interconnection www.pjm.com/index.jsp
10. Grid Florida (no website available)
11. Alberta RTO www.aeso.ca
12. Ontario Independent Market Operator www.theimo.com

Regional Reliability Councils

· Listing is approximately West to East, North to South

1. WECC (Western Electricity Coordinating Council) www.wecc.biz
2. MAPP (Mid-Continent Area Power Pool) www.mapp.org
3. SPP (Southwest Power Pool, Inc.) www.spp.org
4. ERCOT (Electric Reliability Council of Texas, Inc.) www.ercot.com
5. MAIN (Mid-America Interconnected Network, Inc.) www.maininc.org
6. ECAR (East Central Area Reliability Coordination Agreement) www.ecar.org
7. SERC (Southeastern Electric Reliability Council) www.serc1.org
8. FRCC (Florida Reliability Coordinating Council) www.frcc.com
9. NPCC (Northeast Power Coordinating Council) www.npcc.org
10. MAAC (Mid-Atlantic Area Council) www.maac-rc.org

Bibliography

Utility Operations, Quality Control, and Grids

1. Power System Operation, Second Edition, by Robert Miller, 1983. McGraw-Hill
2. Electrical Power Systems Quality, by Dugan, McGranaghan, and Beaty, 1996. McGraw-Hill
3. Be Your Own Power Company, by David Morris, 1983. Rodale Press
4. Energy Isn't Easy, by Norman Smith, 1984. Publisher: Coward-McCann, Inc.
5. Energy: New Shapes, New Careers, by Reed Millard, 1982. Julian Messner.
6. Reinventing Electric Utilities, by Ed Smeloff & Peter Asmus, 1997. Island Press
7. Electric Power Generation Association, www.epga.org
8. Federal Energy Regulatory Commission (FERC) , www.ferc.gov
9. North American Electrical Reliability Council (NERC) www.nerc.com
10. Electric Reliability Council of Texas, Inc. (ERCOT) www.ercot.com/Index.htm
11. California ISO (CAISO) www.caiso.com
12. The California Electricity Crisis by James L. Sweeney, 2002. Hoover Press
13. "Power Surge" by Rob Wherry, *Forbes Magazine*, August 15, 2005.

Department of Energy (DOE) Related Sites

1. Department of Energy (DOE) www.energy.gov
2. Energy Information Administration (EIA) www.eia.doe.gov
3. [Office of] Efficiency and Renewable Energy (EERE) www.eere.energy.gov
4. Office of Fossil Energy (in Dept of Energy) www.fossil.energy.gov
5. Electric Transmission and Distribution Office www.electricity.doe.gov
6. Science (Office of Science) www.sc.doe.gov
7. Nuclear Regulatory Commission (NRC) www.nrc.gov
8. Civilian Radioactive Waste Management (OCRWM) www.ocrwm.doe.gov
9. Yucca Mountain Project www.ocrwm.doe.gov/ymp/about/index.shtml
10. International Nuclear Safety Program http://insp.pnl.gov
11. International Nuclear Safety Center, Argonne Laboratory www.insc.anl.gov
12. National Energy Technology Laboratory (NETL) www.netl.doe.gov
13. National Renewable Energy Laboratory (NREL) www.nrel.gov
14. Oak Ridge National Laboratory www.ornl.gov
15. Los Alamos National Laboratory (LANL) www.lanl.gov/worldview
16. Pacific Northwest National Laboratory (PNL) www.pnl.gov
17. Starlight, from PNNL/DOE http://starlight.pnl.gov

Index for Utilities and Grids

www.ingramcontent.com/pod-product-compliance
Lightning Source LLC
Chambersburg PA
CBHW081238180526
45171CB00005B/468